Where to find ...

GOLD &GEMS IN NEVADA

by James Klein

Gem Guides Book Company
315 Cloverleaf Dr., Suite F
Baldwin Park, California 91706

Front and back cover artwork by Albert Jamison
Photos courtesy of Stan Paher collection.

ACKNOWLEDGEMENTS

My thanks to the Nevada State Bureau of Mines
and the Mackay School of Mines for their help and material.

Note: Physical hazards may be encountered in visiting areas described in WHERE
TO FIND GOLD AND GEMS IN NEVADA, particularly old mining locali-
ties. Readers should take proper precautions. Due to the possibility of per-
sonal error, misinterpretation of information and the many changes of land
ownership and road conditions over the years, author and publisher assume
no responsibility for any loss, damage or injury to any individual or group
using this publication.

OTHER WORKS BY THE AUTHOR

———————

WHERE TO FIND GOLD IN SOUTHERN CALIFORNIA

WHERE TO FIND GOLD IN THE DESERT

WHERE TO FIND GOLD IN THE MOTHER LODE

DRYWASHING FOR GOLD

JAMES KLEIN'S MAPS OF WHERE TO FIND GOLD IN
SOUTHERN CALIFORNIA

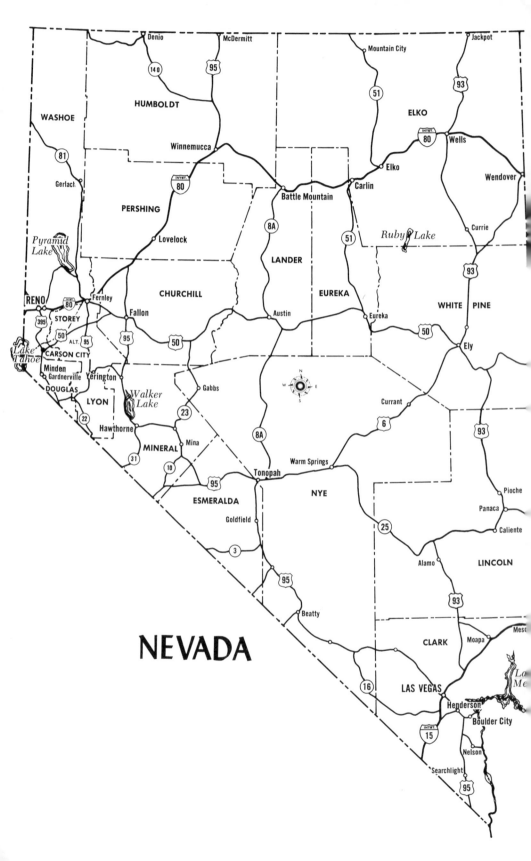

NEVADA

CONTENTS

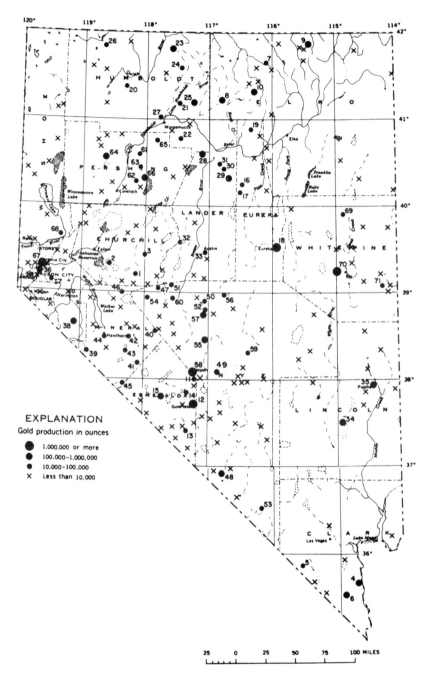

—Gold in Nevada

6

INTRODUCTION

Nevada is like a wild mustang that has been captured but never broken. Sure there are some cities that have managed to get a foothold on a piece of it, but get out of those cities and Nevada is still wild, rugged and beautiful. I have been planning this book for several years but have hesitated to write it because of that reason. It is hoped that those of you who discover Nevada through this book respect it. In many areas things are as they were a hundred years ago. We have seen shacks that still have their furnishings in them, and stores with ancient canned goods still on the shelves and mining carts still on their tracks in the mines. We left them as we found them so that others could enjoy them, too. I hope you will do the same.

I know you are going to enjoy exploring Nevada, and there is a good chance you will be rewarded with some good finds. There is a moment I'll always remember from one of my prospecting trips into the Nevada back country. My son Andy and I had come upon a deserted town and were just looking around. We drifted apart and after a few minutes Andy let out a war-cry and came running to me. He was clutching a bottle he had found in the dump behind one of the old houses. It was a small medicine bottle dating from the late 1800's. It was in beautiful condition. He still has that bottle sitting on a shelf in his room, and I still have the memory of a 14-year-old boy's excitement and joy in discovering a bit of our past. I hope you are as lucky as I am....

HISTORY OF GOLD IN NEVADA

The earliest known placer gold mining in Nevada is thought to have taken place in Tule Canyon. This was prior to 1848, and was done by miners from Mexico. They are said to have taken out more than a million dollars in gold. State geologists feel that this is too high a figure. Either way the evidence is too meager to really tell for sure. All we can say is that there was gold mined in Nevada before the rush to the Mother Lode in California.

The first recorded discovery of gold in Nevada was made by Abner Blackburn in July of 1849. He found his gold in what is now known as Gold Canyon below Virginia City. There was no great excitement caused by his discovery. Everyone was too fired up over the huge finds being made in the El Dorado of the Mother Lode, in the Sierra Nevadas of California, to think of anything else. It was only after being disappointed in their quest on the western slope of the Sierras that they returned to Nevada. Even then, it is estimated that only about 200 men were working the placers in Gold Canyon in the years 1849 to 1857. Compared to the thousands working the canyons of the Mother Lode, this was very small. There were other things that made the Nevada diggings less attractive to the miners. One was the harshness of the climate and the starkness of the countryside. Another was "that damn blue stuff" that kept plugging up their riffles. That damn blue stuff would later make millionaires out of many of them—it was silver ore. Due to the silver alloy Nevada gold also brought less per ounce than the California gold. As the miners worked their way up the canyons following the placers, the gold value dropped even more. In the beginning, Gold Canyon dust had brought $15 to $16 per ounce in Hangtown. It dropped to $11 an ounce as it was found to contain more of the blue alloy.

The "blue stuff" plugged up their rockers and sluice boxes and filled up their pans. Things got so bad that at one time in late 1850, when the water had ceased to flow till spring and food and supplies had gotten so high that the miners couldn't afford them, a relief party had to be rushed over from California.

The first camp or town to spring up due to the placers was called Johntown and was about four miles up the canyon from Stafford Hall's Station. The first actual settlement in Gold Canyon was in the valleys and was named Mormon Station. At first the Mormons devoted themselves to farming, but as the diggings grew richer and richer the Mormons left their farms and joined the miners in reaping the gold. The Mormons had brought in about a hundred orientals to help them build an irrigation ditch from the Carson to their farms, but they also got gold fever and joined in the rush. The orientals built their cabins around Stafford Hall's Station and that area became known as Chinatown. This area soon became the center of the wild life for the miners. Here they could drink and gamble away all the gold they could dig in a day.

8

Early in the boom Mexican miners had tried to tell the prospectors that for every dollar they kept, they were throwing away two. The Mexicans, some of whom had worked in the silver mines of Sonora, knew what that damn blue stuff really was—rich silver ore.

In 1857, a number of prospectors who followed that urge that seems to dwell deep in every gold seeker's soul—the lure of finding that big one, that one discovery that will make them rich forever—wandered into Six-Mile Canyon, which is the next ravine two miles north of Gold Canyon. Here the gold was not found in the gravels, but in a blue clay that had to be disintegrated in water to release the gold. The miners were able to recover an average of an ounce a day but were puzzled by the peculiar characteristics of the placer gold. As they worked their way up both ravines, the gold miners struck rich pay dirt in the decomposed outcroppings of the prominent quartz ledges crossing the head of both ravines. It was not until 1859 that the miners discovered why the gold they had been washing out of the alluvium below the quartz ledges was worth so much less than the gold found in the placers on the western slope of the Sierras. The decomposed quartz, almost black in color, was found to be rich in silver in combination with the gold. This was the beginning of mining on the Comstock Lode and the discovery of the first noted silver mining district in the United States. Even with the large amounts of silver the early miners made more from the gold content of the ore due to the large difference in price of an ounce of gold to an ounce of silver.

After the first rush to the Mother Lode, the miners who didn't make it there began spreading over Nevada, going not only to Gold Canyon but to other parts of the state. In fact, gold is found in every county excepting Lincoln. The prospectors wandered from place to place following every report of rich diggings being discovered. Many placer deposits and lode deposits were found and worked during this period. From the earliest days gold mining in Nevada has seemed to go in cycles; whenever a strike of major importance occurred, a new rush of excitement brought gold seekers rushing to the area. After a while things would quiet down, then boom, someone would make a rich strike and the excitement would begin all over again. Even today this is true. I have seen millions of dollars being spent in places such as El Dorado just to work the tailings from the early days, and I know people who are working their own mine by hand in places like Manhattan and doing quite well. I have also seen thousands of dollars worth of equipment sitting idle and rusting away at the property owned by the Summa Corporation. A lot of mining is going on in Nevada today, and a lot of gold is being mined.

The Chinese played an important part in the location of the placer deposits. Joining the Chinese, who were brought in to build the irrigation ditch and who left to join the gold rush, were the thousands of Chinese who were employed in building the railroads in the state. After the railroads were completed the hard-working Chinese headed for the placer mines, mostly in the northern portion of the state. They did so well that a large number were brought over by their fellow countrymen just to work the placers. There was a great deal of prejudice against them and they were discriminated against in the early laws of the state. We owe a debt to them as they were the first to locate many of the deposits we work today. Most of the Chinese placer miners had left the state by 1900.

The amount of placer gold recovered in the state prior to 1900 will never be known for sure as no accurate records were kept. The Chinese in particular were very secretive about the amount of gold they recovered, and much of the gold they mined never made its way to the mints or to the Wells Fargo office. There are some estimates given for some of the districts prior to 1900. Keep in mind that these figures are calculated at the price of gold being around $20 an ounce. American Canyon and Spring Valley in the Humboldt Range are said to have produced over $10,000,000; Barbara and Wright Canyons in the Sierra district produced around $2,000,000; and the Osceola district produced $3,000,000. The placers at Tuscarora are known to be around $7,000,000. Again, all these figures are at the low price of gold at that time. Just compare those figures with the price of gold today. Since 1900, some of the most productive years were from 1911 to 1922 when over 160,000 ounces of placer gold were mined. The most productive year during this period was 1915 when more than 19,000 ounces of gold were mined by placer methods. More gold has come from the lode deposits and as a by-product of such ores as copper and silver. Total gold production for the years 1903 to 1940 give you a better idea as to the large amounts of gold that have been recovered, making Nevada the 5th largest gold producing state in the union. During this period 12,730,729 ounces of gold were mined in the state. The counties producing the largest amounts were Esmeralda with 4,863,958 ounces; Nye, with 2,849,397 ounces; and White Pine with 1,196,805 ounces. The gold mines were closed by government order during the Second World War and there was little production during this period. Since the war there have been peaks and valleys in looking at the charts. The peaks have been due to large scale placer operations at Round Mountain, large-scale open pit mining at the Goldacres pit in the Bullion district, and at the Getchell Mine, in addition to significant sustained amounts of by-product gold from Ely. The more recent boom in gold production has been led by the Carlin Mine. The Carlin Mine is operated as a subsidiary of the Newmont Mining Co. which also operates the Blue Star Mine nearby. The Carlin is an open pit mine where the gold is recovered by cyanide process. They are running 2,500 tons of ore per day. In 1979 they processed 843,700 short tons of ore. The ore ran 0.186 ounces of gold per ton. In all, they mined 133,600 troy ounces of gold in 1979 at the Carlin.

There is still gold to be found in Nevada and today's modern prospector is searching the hills and gullies for it, and finding it.

NOTE

Each chapter begins with some general information on what you will find in the county. The gold-bearing regions are always the first listings in the chapter and are in alphabetical order. They are followed by selected gem and rockhound areas, though not every gem collecting area is listed—only the ones of most interest to the general public. See pages 52-53 for map and table on Nevada gems.

After the gem areas you will find some ghost towns listed. Again, not every ghost town or site is listed, mainly the ones that offer good possibilities for the metal detector and camera user and that can be reached fairly easily.

At the end of each chapter we will have one or two of the better treasure tales based on the history of the county for all of you who, like me, still believe it is possible to find that pot of gold at the end of the rainbow.

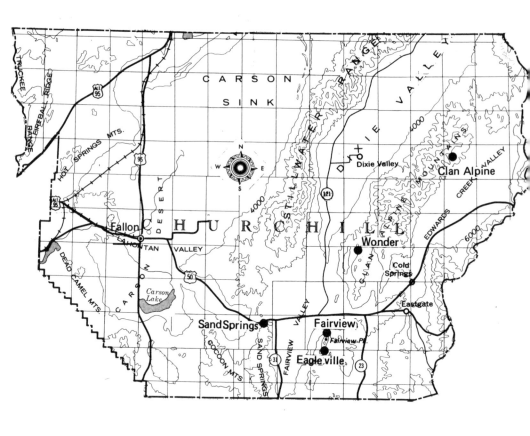

12

WHERE TO FIND GOLD AND GEMS

Churchill County

Churchill County has one fairly good placering region and several areas where large amounts of gold have been recovered as a by-product of silver mining. For the gem hunter banded rhyolite, agate and jasper can be found here. There are also a couple of lost treasure stories and some ghost towns for the treasure hunter.

Eagleville Placers

The Eagleville district is also known as the Hot Springs district and is located a short distance south of the old mining camp of Fairview near the county line in a canyon south of the Eagleville Mine. To reach the area take Highway 50 out of Fallon, then go south on Highway 23. Fairview is a ghost town and is located almost on the old Pony Express Route. The first placers were worked around the turn of the century and activity began again during the depression years. The gold is coarse and finely divided and the gravels are angular so the gold has not traveled too far from the source.

Fairview District

The gold from this district is a by-product of silver mining and 53,100 ounces have been recorded. The ore occurs in fissure veins in Tertiary andesite. Little has been done here recently except some prospecting. The district is in the area around Fairview Peak, 42 miles southeast of Fallon.

Sand Springs District

This district is located about 25 miles south of Fallon, near Highway 50 before you come to the junction of Highway 31. The gold was a by-product of the silver ores. Recorded production is 20,875 ounces. You can find a little color in some of the canyons here.

13

Wonder District

This district is the largest producer with 73,890 ounces of gold being recovered as a by-product of silver mining. The veins are in rhyolite and very rich in silver. The mines are located on the western slope of the Clan Alpine Mountains. Take Route 121 north about 6 miles east of Frenchman off Highway 50 to reach the region.

Some minor production has been reported in the Terrill Mountains and Baenett Hills south of Fallon. There has also been some placer gold mined near the ghost town of Jessup, which is located north of Highway 95 near the Pershing County line.

Banded Rhyolite

Banded rhyolite can be found in a gravel pit by going east on Highway 50 for 11.5 miles from Fallon to a turnoff, then going 2.5 miles east to the site. About 1 mile past the gravel pit you can also find some green agate as well as lace agate.

Jasper-Agate Nodules

There are some jasper-agate nodules found on the hillsides near the U.S. Navy gunnery range near the junction of Highway 50 with Highway 31. Take the Rawhide-Sheelite turnoff from Highway 50 and go 7 miles south to a road going southeast. Follow this for about 8 miles to a branch road heading east. The nodules are in the canyon at the end of the road.

Clan Alpine Ghost Town

There are still some ruins of the company office buildings standing here as well as foundations of other buildings. The town was short-lived, 1866-1867, but you might find some artifacts with your metal detector. Go east on Highway 50, 14.2 miles from junction with Highway 2 east of Frenchman to a turnoff on your left. Go 3.9 miles on this road to another turnoff on your left, then take this road about 1 mile to Clan Alpine.

Clark County

Not all the gold in Clark County has been mined by the casinos in Las Vegas. Gold production estimates go as high as one-half million ounces. There's an old fellow who has his own private museum in Nelson and he has been prospecting the area for over 40 years. He might give you a tip or two if you asked. There is quite a bit of activity going on in this area. They are re-working the tailings which contain an ounce of gold per ton according to the people running the operation. A well-known treasure hunter friend of mine found a gold coin with his metal detector at an old stage stop near here recently.

Eldorado District

This district is also known as the Colorado or Nelson district. It is in the Opal Mountains near the Colorado River. Eldorado Canyon drains into the river. There have been some deaths due to drowning here during high water and the road to the river may be closed at times. Production up to 1964 was 101,729 ounces of gold from lode deposits and as a by-product of silver mining. There have been conflicting estimates as to the amount of placer gold mined here. Every type of placer mining has been tried here. Dry-washing in the canyon produced gold in the 1890's. Several shafts have been sunk in the canyon in attempts to hit pay gravel. Around 1909 two dredges were set up to work the bars of the Colorado River but the results were not too good. There is gold to be found in the bars but it is very fine and distribution is erratic. To reach this area take Highway 60 off of Highway 95 south of Boulder City.

Moapa District

You can get a little color along the Muddy River about 3 miles southwest of the town of Moapa. The gold is found in sand heavily stained with iron oxide. No figures are available on the amount of gold produced here.

Gold Butte District

This district is located in the south end of the Virgin Range of mountains on the north shore of Lake Mead. The best placering area is now under the waters of the lake and is known as Temple Bar. Some dredging has been done here with fair results. The deposits were first worked by drywashing in the 1920's. The placer deposits cover a large area and the gravels have depth of from 2 to 20 feet. The area just above bedrock is the most productive. The gravels are cemented and the gold is very fine.

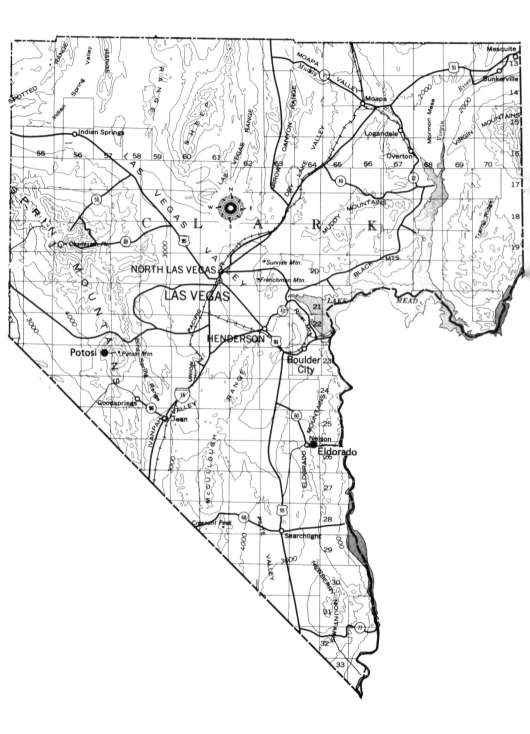

Goodsprings District

This lode gold mining district is located in the Spring Mountains. The gold occurs in pyritic fracture fillings in or near granitic dikes. Recorded production is 58,815 ounces. The Potosi Mine here is the oldest mine in the state. To reach Goodsprings take Highway 15 to Jean, then take Highway 53. It is about 7 miles to the town.

Searchlight District

This is the largest producing district in the county with 246,997 ounces of gold reported. The ore is in gneiss near contact with Tertiary quartz monzonite body. The two most productive mines were Duplex and the Quartette. Some placer gold has been found here, but all the recorded production is from the lode deposits. Searchlight is located on Highway 95, 36 miles south of Boulder City. The town was formed in 1897 as a mining camp and is a good location for metal detecting.

Turquoise

Clark County has been one of the largest producers of turquoise in the state. The Sullivan Mine is located about 4 miles east of Boulder City on the south side of Highway 93. There are quite a few holes dug here as well as two shafts. The Simmons or Crescent Peak Mine is about 12 miles west of Searchlight off the road to Nipton. The deposits are on the south and west flanks of Crescent Peak. There are two patented claims, and quite a bit of work has been done here as you can see by the different pits, shafts and tunnels. This mine was worked by an ancient people at least 200 years before Columbus discovered America. Artifacts and ruins revealed a very sophisticated operation. It is thought the mine may have been operated by Aztecs or Toltecs, as they cut and polished the turquoise on the site after mining it and understood the economy of transporting finished stones rather than rough ore, with its waste rock.

Another mine in this area was the Morgan which was located 3 miles west of Crescent Peak and 1 mile south of the old town of Crescent.

Petrified Wood

Take Highway 15 north from Las Vegas 22 miles to Crystal, turn off on Highway 40 (Valley of Fire State Park turnoff), and stay on 40 past the fork (about 3-3.5 miles). Petrified wood can be found alongside the road up to the State Park boundary. Pink opal and red jasper can be found by taking the fork to the right going 5 miles toward the Muddy Mountains to the entrance of a canyon. The red jasper and pink opal nodules are about 4.5 miles from there.

Potosi Ghost Town

The town of Potosi is known to have been in existence as early as 1855 when the Mormons were mining what they thought were lead deposits. To reach Potosi take Highway 15 south out of Las Vegas to the highway going to Pahrump. Follow this road for about 18.8 miles and you will come to a road on your left; the site of Potosi is 4.4 miles down this road.

Treasure Tales

Somewhere near the town of Mountain Springs two chests filled with silver are said to be buried. In 1897, two brothers and a driver were bringing a load of supplies from Utah to the settlement at Mountain Springs, and also two chests containing $20,000 in silver they were going to use to invest in the mining company there. Right before they got to the camp they were attacked by Indians. One brother was killed in the fight and the other brother and the driver were wounded so badly that they were left for dead. It was several days before they were found. They were taken to the camp of the Colorado Mining Company and before they died told the person who was nursing them that they had buried the two chests of silver behind a large rock sometime during or after the attack. A search was made after the two died from their wounds and exposure. The burned out remains of their wagon were found but the searchers were unable to find the chests. A good metal detector and a little searching might make someone rich here.

Golden Gem Mill near
Nevada-Arizona border.
Photo 1978

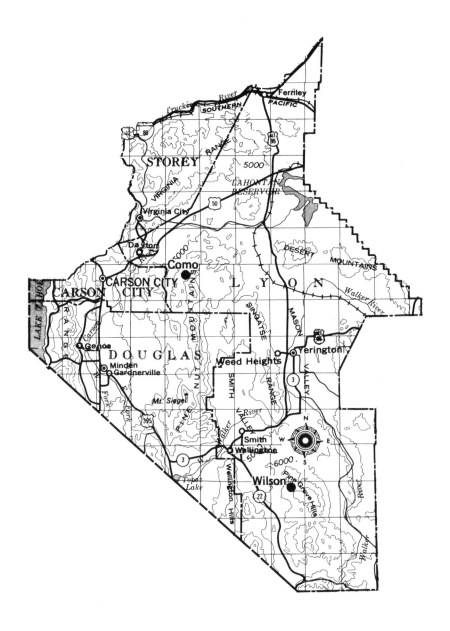

Douglas County

Some gold was mined here in the past, but there is no recorded production. Some estimate the amount of gold mined here at around 10,000 ounces, most of which came from placer deposits. There is some turquoise, as well as jasper and agate found here. One of our most recent treasure stories also takes place here. If you want you can come down out of the mountains and try your luck at the gaming tables in Stateline or just dip your feet in the cool waters of Lake Tahoe. The Buckskin gold district is shared with Lyon County and is reported in that chapter.

Genoa District

The Genoa district is located on the eastern slope of the Sierra Nevadas in Douglas County, just west of the old town of Genoa. Mining was first started here as early as 1860. A lot of work was done trying to develop the lode deposits but with little success. The Tertiary gravels here contain small amounts of gold.

Mount Siegal District

There has been quite a bit of gold mined here and some large nuggets also found here. The district is located on the north side of Mount Siegal in the Pine Nut Range of mountains about 20 miles east of Minden and 65 miles southeast of Reno. Due to the lack of water the placers were worked only in the spring when there was water from the melting snow in the early days. The placer deposits occur in a large depression in granite mass which makes up the range. This depression has been cut by recent water courses and the gravels consist of some large boulders along with rocks, sand and soil. The gold is scattered over a large area with some concentration in the ravines or on a false bedrock of pipe clay. The source of the gold is an ancient Tertiary river and from erosion of gold bearing quartz veins and stringers in the mountains. Buckeye Creek has been very productive.

Turquoise

To reach the turquoise deposits at Buckskin Mine take the road along Colony Ditch 1 mile west of Wellington for 7.3 miles. Turn west, then go north for 8 miles. At the north end of Artesia Lake go east and northeast past the ranch for about 4 miles to a road heading due north. Take this road to the mine; it is less than a mile.

Agate and Jasper

Agate and jasper can be found in the Pine Nut Mountains around Mount Siegal.

Treasure Tales

Harrah's Lake Tahoe Resort Hotel in Stateline was robbed of $178,500 in 1972. Most of the money was recovered by the F.B.I. and the four men who committed the robbery were arrested and sent to jail. They have refused to tell the hiding place of the rest of the loot; $52,220 is hidden. Most people feel it is hidden somewhere near Stateline.

Elko County

There are quite a few locations here for the gold prospector, both placer and lode. Some of the placer gold is coarse and makes excellent pieces for jewelry making. You can find turquoise, cinnabar, agate, quartz crystals, petrified wood, beryl and other stones here also. For the treasure hunter there are several ghost towns and some lost treasure stories.

Alder District

This district is located in north central Elko County east of Wild Horse Reservoir. There is a good gravel road off Highway 51 leading to this area. The area is easy to find, just look for the tailing piles alongside the creek. Quite a bit of work was done here in the early days and you can still get a little color today.

Aura District

Also known as the Bull Run, Centennial, or Columbia this district is located in the central part of the Bull Run Mountains. The ghost town of Aura site is reached by taking Road 11A north off Highway 11. Most of the gold was found in Sheridan, Blue Jacket, and Columbia Creeks. The placer deposits were discovered in 1869 and worked mostly in the 1870's. The lode deposits were more productive than the placers.

Charleston District

The Charleston district is also known as the Copper Mountain or Cornwall district. It is located on the south side of Copper Mountain in the Jarbridge Range. Placer gold was first found here in 1876 on 76 Creek. Quite a bit of gold was recovered from 76 Creek since there was more water here than in the other placer locations in the district. Other productive areas are Pennsylvania Gulch, Dry Ravine, Badger Creek, and Union Gulch. The placers extended for miles along the Bruneau River as well. The gravels are made up of well-rounded pebbles mostly coarsely crystalline rhyolite, and are 50 feet thick in some places.

Edgemont District

Little has been done here since around the turn of the century. Total production is put at about 48,500 ounces of gold all from lode deposits. The veins are narrow and occur in cracks or grooves in contorted and fractured quartzite. This district is located on the west side of the Centennial Mountains in the northern portion of the county.

Gold Basin District

Also known as the Rowland district it is located in north central Elko County. Some placer gold was recovered here on the north fork of the Bruneau River, but no appreciable amounts of gold have come from this district.

Gold Circle District

The Gold Circle district is also known as the Midas district. The ghost town of Midas can be reached by taking Highway 18 west off Highway 51. It is 49 miles from the junction with 51 on 18 to reach the site. 109,765 ounces of gold from the lode mines have been recorded here. There has also been a small amount of placer gold found here as well. The lode deposits are veins in shear zones in the older of two Tertiary rhyolite flows.

Island Mountain District

This is also known as the Gold Creek district and can be reached by taking the road to the Gold Creek Ranger Station. The placers are about halfway between the Ranger Station and Highway 51. The district gets its name from an isolated mountain that rises over 1,000 feet above the surrounding terrain. The average elevation is high, nearly 8,000 feet above sea level. The placers were first worked in 1873 and quite a bit of gold has been recovered. Some of the more productive areas have been in Hammond and Coleman canyons, and in Hope Gulch. Early reports run as high as $2.50 per pan which would be about $100.00 per pan today. The gold occurs in gravels near the surface and is fairly coarse. The source of the gold is thought to be in quartz veins in the area and at the head of Gold Creek.

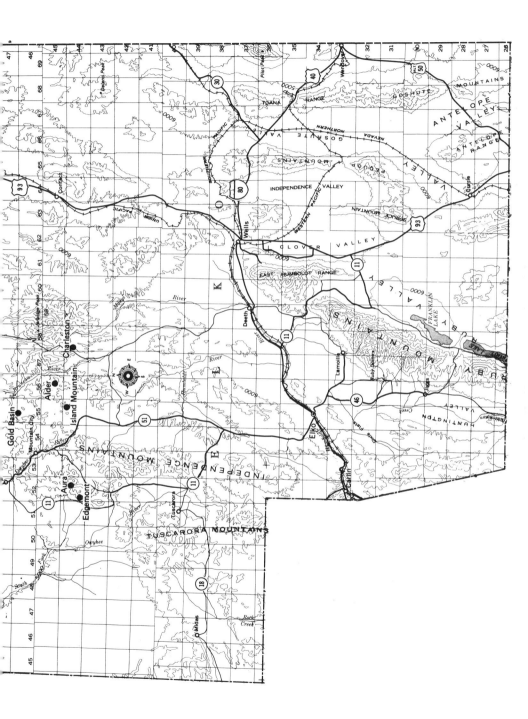

Jarbridge District

The old mining camp of Jarbridge still lives today, maybe not as loud or as rowdy, but it is not a ghost. It is located in what is now the Humbolt National Forest. The easiest way to reach the location is to take Highway 51 from Elko and cut across; or you can take the road north out of Deeth. Lode mining was the largest source of gold here with 217,800 ounces having been recorded. Some placering has been done along the Jarbridge River with little success. The ore is gold-bearing fissure veins in the older of two Tertiary deposits, and the most valuable deposits are in veins that strike northeast.

Mountain City District

This is also known as the Cope district and is located in the northeastern portion of the Centennial Range of mountains on the Owyhee River. The old mining town of Mountain City can be reached by taking Highway 51 north out of Elko. Mountain City is about 85 miles from Elko. Both placer and lode gold can be found here. There are no recorded production figures for this district. Grasshopper Gulch, north of Sugar Loaf Peak, was the most productive placering area along with Hansen Gulch, which is a tributary to Grasshopper Gulch. The area along the river north of the city was also productive.

Tuscarora District

The old mining camp of Tuscarora can be reached by taking Highway 51 north out of Elko for 27 miles to the junction with Highway 11. Take Highway 11 for 18 miles to the Highway 18 turnoff. Follow 18 for 7 miles to Tuscarora. The district is located on the southeastern slope of Mount Blitzen in the Independence Range of mountains. Production is set at 100,000 ounces at least, lode and placer. The placers run for nearly 3 miles in gravels four to ten feet deep. The ore is quartz and quartz stringer with gold. The placers are rather narrow and rich in spots. The largest nugget found here weighed nine pounds and one-ounce nuggets were common. The gold is found mostly on bedrock or near bedrock. The placer workings are confined to the gullies on the slopes of the hills bordering the west side of Independence Valley. Some of the more productive areas were Eureka Gulch, Gardner Ravine, and Stovepipe Gulch. This district was worked heavily by the Chinese. The placers are fed by the veins that occur in rhyolite and andesite north and west of the diggings.

Remains of early gold mine circa 1900 - northern Nevada.

Van Duzer District

Van Duzer Creek is a small stream that flows eastward and joins the Owyhee River about 6 miles south of Mountain City. The placers are gold-bearing gravels in and along the creek bed. Estimated production is set at $100,000. The deposits occur all along the creek for several miles. Most of the work done here has been on a small scale, but hydraulic mining was done in the 1890's. Nuggets weighing five and six ounces have been found here. The source of the gold is thought to be from quartz veins at the head of the stream.

Turquoise

Turquoise can be found at the Stampede Mine, which is located about 9.5 miles southeast of Tuscarora. It is an open pit mine and the turquoise occurs as nodules and seams in black chert and gray quartzite. It is solid blue, matrix-marked, or spider web. All three varieties are also found in float nearby.

Agate

Agate can be recovered on Jack Creek north of the town of Jarbridge. Take the road north out of Jarbridge about 3.5 miles to the creek. Take the trail east into Jack Canyon to the fork where you will find a seam agate deposit.

Beryl-Garnet-Quartz Crystals

You can find beryl, garnet, galena, and quartz crystals in the Ruby Mountains. Take Highway 46 south out of Elko 33 miles to the town of Jiggs. Three miles past Jiggs you come to a road heading east; this is the Harrison Pass road. Follow this until you come to a junction (about 15 miles), turn left and go 3 miles to Ruby Valley. The best location is around Smith Creek.

Cinnabar in Opal

Some good specimens have come from the Rand Mine in Midas. See the Gold Circle district for directions to Midas.

Cinnabar in Opalite

Fourteen miles west of Tuscarora take the road on your left. Follow this road for around 20 miles to the Silver Cloud Mine. Look in the washes and dumps. There is also agate, chalcedony and petrified wood in this region.

Pink Jasper and Agate, Petrified Wood

Texas Spring Canyon is well known for its good quality specimens of pink jasper and pink agate as well as petrified wood and limb casts. There is a road going east after you pass Knowl Creek that leads you into the area or you can take the road going east off 93 going to the San Jacinto Ranch south of Jackpot. It is about 15 miles from Highway 93 when you take the San Jacinto Road.

Rio Tinto Ghost Town

The ghost town of Rio Tinto can be reached by taking Highway 51 south from Mountain City 2.1 miles to a road going west; Rio Tinto is about 1 mile down this road. Rio Tinto was founded in 1919 and deserted after the Second World War. There are some good ruins here and it is a good metal detector site.

Treasure Tales

There is a fairly recent treasure story about a location near Jarbridge. The tale deals with what is thought to be the last stage coach robbery in the West. Most of the facts can be checked, which is not the case in most treasure stories. The story starts on December 5, 1916. Even in those days people still looked forward to the arrival of the stage. This day a small group of town-folk had gathered at the stage stop to wait even though it was bitter cold and snowing. When the stage got later and later a few of the men set out to look for it, fearing there might have been an accident on the slippery road into Jarbridge Canyon. They found the stage only about a quarter of a mile from town. It had been pulled off the road and hidden in a clump of trees. Suspicion fell on a miner who was not well liked in the camp. His name was Kuhl and he was soon arrested and charged with murdering the driver, who was found shot to death sitting on the driver's seat of the stage, and with stealing $4,000 in cash that had been in the mail pouch. He was convicted and sent to jail for life in 1917. He was released in 1945, but never returned to Jarbridge. The $4,000 was never recovered and people still feel it is buried some place in the canyon.

There are several lost mine stories in Elko County. One is located in the Harrison Pass area south of Elko in the Ruby Mountains. The Spanish are said to have worked a rich gold mine there but no one has ever been able to locate it. Another story revolves around a rich gold silver vein found by a sheepherder on the Twin-Bridges Ranch near Elko.

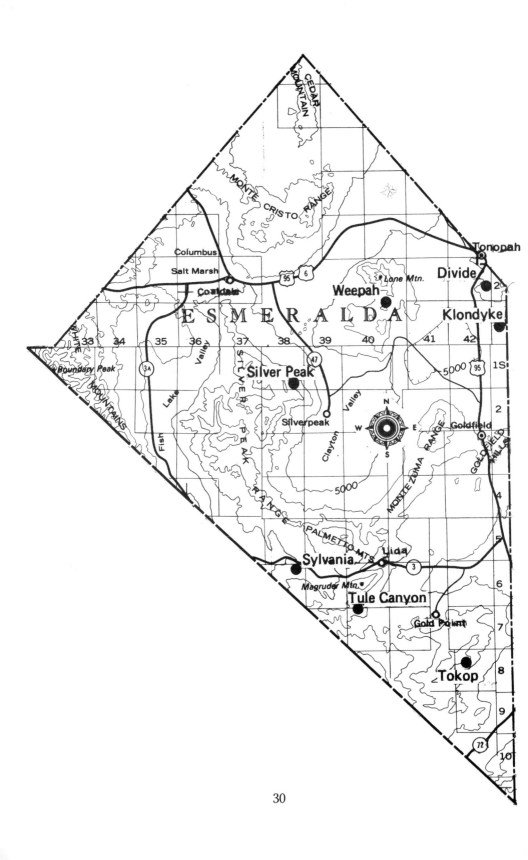

Esmeralda County

When you think of gold in Esmeralda County, the first thing that comes to mind is Goldfield, but the placers here were worked long before the strike at Goldfield. Some experts think the placers were being worked before the rush to California's Mother Lode. It is known that the placers in Tokop and other districts were being worked in the 1860's. The Goldfield district has produced 80% of the county's more than 100 million dollars production. Silver Peak and Divide have been the other most productive regions. There are several excellent gem collecting areas also in the county as well as several ghost towns. I had a funny experience here. My son and I were on a prospecting trip and coming from California we were not familiar with all the local customs. We had been poking around near Gold Point and had decided to move to a new location. We had just reached the highway when we saw a sign that said "cold drinks." It was hot, and a cold Coke sounded good so we stopped. We were in a hurry, so I said that I would run in quickly and get a couple for us. Andy who was 14 waited in the car. Well, the first thing that didn't seem right to me was the fact that the door was locked. There was a sign that said "ring bell;" I figured that being in an out-of-the-way place maybe the owner was just being cautious. I rang the bell and in a moment or two it was opened by this kinda middle-aged lady. I told her we wanted a couple of those cold Cokes. She gave me a big smile and said, "Come on in, honey." Inside was a nice big bar but no other customers. About the time I got a few feet inside the lady rang a bell. "All right, girls, we have a customer." Girls came from all directions and lined up in a straight line. "Take your pick, mister," she said. I guess I musta stood there quite a while with my mouth open. The girls were not over dressed, to say the least. Before I could explain that all I wanted was a couple of Cokes, really, one girl came up and said, "How's $50 sound to you?" Again I repeated that all I wanted was a couple of cold drinks. I guess they thought I wanted to argue the amount because a price war started. It took a while but I finally convinced them that my son was out in the car in the hot sun and all I really wanted was something cold to drink and was able to escape. That's how I learned that those red lights you see shining on those buildings out there in Nevada don't mean Fire Department.

Divide District

All the gold in this district (26,500 ounces) is a by-product of the silver mines. The most productive year in this district was in 1922 when the mines recovered more than one-half million dollars in gold and silver at the old prices. This district is on Gold Mountain just south of the town of Tonopah.

Gilbert District

This district is also known as the Desert district. It is located on the north slopes of the Monte Cristo Mountains in the northwest portion of the county near the border with Mineral County. There have been reports of small amounts of placer gold found here as well as lode deposits. The most productive period was during the 1920's.

Goldfield District

Since Goldfield was one of the last sites of a gold rush in this century, it is well documented and many of the great stone buildings still remain. You get the feeling here that Goldfield is only sleeping and it is only a matter of time when it will be booming again. The veins were first discovered in 1902, but production did not start until the end of 1903. By 1906 mining was in full swing and the town was founded. Production continued to grow and reached its peak in 1910 when it produced $10,711,664 for its owners. Production stayed high until 1919, then it had its ups and downs until 1940 when the mines were closed by government order. Goldfield has produced 4,195,000 ounces of gold, all from lode deposits, some of which were extremely rich. The town of Goldfield is located on Highway 95 about 25 miles south of Tonopah and well worth a trip.

Gold Point District

The original name of this district was Hornsilver. Recorded production is 25,000 ounces all from lode deposits. The veins occur in calcareous shale near a granite intrusive. This is a well-preserved camp and some people still live here. Mining began here in 1908 and reached its high point during the years 1915, 1916, and 1917. There have been small amounts of placer gold found near here but no mining of the placers has proven worthwhile. To reach Gold Point take Highway 95 south from Goldfield for 15 miles to Highway 3. Follow Highway 3 west for 7 miles to Highway 71 (gravel). It is 7 miles to Gold Point on this road.

Klondyke District

The Klondyke district is located about 14 miles south of Tonopah in the southern Klondyke Hills near the border of Nye County. About $30,000 is the best estimate of the production here. The Chinese are said to have mined the placers here in the 1870's. The lode deposits were discovered in 1899, and some placering was done here for several years, all by drywashing. One nugget said to have been worth $1,200 was reported found here; at today's prices it would be worth a great deal more.

Silver Peak District

Also known as the Red Mountain and Mineral Ridge districts, it can be reached by taking Highway 6 west out of Tonopah for 35 miles to Highway 47; it is 21 miles to Silver Peak. A 10 stamp mill was said to be operating here as early as 1864. The first recorded production was in 1873. The peak years were from 1908 to 1914, and again from 1933 to 1940. There is still some activity today. The largest single producer was the Pittsburg Silver Peak Mining Company which recorded over $5,000,000. Total production is set at 568,000 ounces, all from lode.

Sylvania District

This is also known as the Pigeon Springs or Palmetto district. The lode deposits were first discovered in 1866 and were the most important. The placer deposits covered a wide area and are found on the eastern slopes of the Sylvania Mountains. The area can be reached by taking Highway 3 west off Highway 95. Eleven miles west of Lida take a dirt road going southwest. Follow this road for about 5 miles to the area of the mines.

Tokop District

Sometimes called the Gold Mountain or Oriental district it is located about 29 miles south of Goldfield. It can be reached by taking the dirt road southeast out of Gold Point for about 4 miles. It was first discovered in 1866 and the rush started in 1871 when rich gold veins were found in the Oriental Mine. Oriental Wash is a small valley between the Slate Range on the north and Gold Mountain on the south. The float was very rich and it was picked and shipped by itself to the smelters.

Tule Canyon District

Tule Canyon district is also known as the Lida, or Alida Valley district. It is located about 10 miles south of the town of Lida at the southern end of the Silver Peak Range between Magrudar and Sylvania Mountains. Placer gold has been found in the district over a ten square mile area. The most productive placer deposits are found in Tule Canyon and gulches feeding into it, like Nugget Gulch at the head of the canyon. The first recorded production is in 1871, but the placers are known to have been worked long before that by miners from Mexico. Production was high from 1873 to 1878 and again from 1935 to 1940. More than one-half million dollars in gold has come from this district. This is a good dry-washing area. Due to the lack of water, the placers have never been completely worked out. The gold is coarse and sometimes still in its quartz matrix. The source of the gold is thought to be the lode veins at the Eagle and other lode mines in the area.

Weepah District

Also known as Lone Mountain, 32,000 ounces of gold have come from this district, all as a by-product of the silver ores. To reach Weepah, go west on Highway 6 from Tonopah about 28 miles to a dirt road on your left. Weepah is 13 miles down this road.

Turquoise

There are quite a few turquoise deposits in Esmeralda County. The area around the old town of Candelaria has produced some good quality turquoise mainly from the Miss Moffet Mine which is about 4.5 miles southeast of Candelaria. The Carl Riek Mine north of the Miss Moffet is also said to have produced some nice stones. The Monte Cristo Range of mountains contains several known deposits. The Carrie Mine has also been known as the Hidden

In the spring of 1927, the mining camp of
Weepah sprang from the sagebrush after two
Tonopah lads discovered $70,000-a-ton ore
in this area. The population of Weepah
reached as many as 2000 before the boom sub-
sided in the fall of 1927. The remains of an
open pit mine and wooden ruins are left to
indicate the site.

Treasure, and the Myers and Bona Mine. It is about 2.5 miles southeast of the old mining town of Gilbert. Some of the turquoise nodules that have come from this mine are equal to those produced in any mine for color and hardness, but the larger stones of high quality are difficult to find. The color of these stones is a beautiful sky-blue. The mine was discovered in 1897 and has been worked off and on since then. Some solid blue and spider webbed nodules have been found at the Monte Cristo Mine.

Other areas in the Monte Cristo Mountains to search for turquoise are: around the Petry Mine which is 11 miles north of Millers, the Blue Bell area on the western slope of the range, on the east side of the plateau near the Royal Blue Mine, which is about a mile north of the Petry Mine, and at the Carr-Lovejoy Mine. The Marguerite Mine is about 3 miles from the Petry Mine and also has produced some nice stones. The Bonnie Blue Mine is near the Columbus Salt Marsh, and the area on the west side of the canyon should be checked.

Turquoise can also be found in the area around Lone Mountain. The Blue Silver Mine is located on the east side of the low hills on the east side of Paymaster Canyon. The Lone Mountain or Blue Jay Mine is about 1 mile east of Paymaster Canyon. The Lone Mountain Mine is famous for its beautiful spider-web turquoise. Farther south in Paymaster Canyon is the Lively Mine; a small amount of high grade turquoise has been found here over the years.

The Smith Black Matrix Mine is located about 0.75 mile east of Klondyke Peak. The turquoise here is found in black chert and the contrast is quite striking.

Apache Tears

Take Highway 6 west from Coaldale 6 miles to Road 3A. Turn left and go 1 mile to a gravel road on your left. Take this road about 5 or 6 miles and look along the side of the road for Apache tears.

Obsidian-Opalite

Take the road north and west from Miller's turnoff, which is 11 miles west of Tonopah. Go about 10 miles to a side road going north. Take this road about 2.5 miles to opalite and obsidian.

Petrified Wood-Agate-Jasper

Take Road 3A south off Highway 6 for 9 miles to a road on your left heading east. Take this road to a sump hole for jasper, agate, and petrified wood. It is about 2.5 miles from 3A.

Goldfield Gem Claim

There is a fee charged for collecting here. To reach the area go north 4 miles from Goldfield on Highway 95. Then go west for 4 miles. You will find good agate, picture chalcedony, green opal, and jasper.

Ghost Towns

Most all of the camps listed in the gold districts are good metal detecting locations. Even though some still have a few inhabitants, most are considered to be ghost towns.

Treasure Tales

Somewhere between Candelaria and Columbus a stage robber buried his loot before being captured. The bandit refused to say where it was hidden and it never has been found. Most of the residents of the area spent a little time searching for it. Take a look around; if you were afraid of being caught with hold-up loot, where would you hide it?

Eureka County

Eureka means "I have found it," and they sure have at the Carlin Mine. There is some confusion about the location of the Carlin Mine. The town of Carlin is in Elko County, but the Carlin Mine is 18 miles northwest of the town in Eureka County. This huge open pit mine is the second largest producer of gold in the country. The Carlin Mine and Homestake Mining Company's big new operation in Napa County, California, prove that there is still a lot of gold to be found. There are several good placering areas here, as well as some turquoise and copper mineral specimens to be found. For the treasure hunter there are ghost towns to metal detect and a treasure story to check out. Eureka County was part of Lander County until 1873 so the early production records are confused.

Buckhorn District

This district is also known as the Mill Canyon district. Almost all of the production of 39,500 ounces has come from one mine, the Buckhorn. The Buckhorn district is located in the southern section of the Cortez Mountains about 5 miles northeast of Cortez, and the mine is located north of the old mining camp of Buckhorn.

Cortez District

48,700 ounces of gold have come from this district, all as a by-product of silver and lead ores. The largest production was from the Garrison Mine which was operated from 1873 to 1896. To reach Cortez, take Highway 21 south from Highway 80 about 20 miles west of Carlin. Go south on Highway 21 about 30 miles to Cortez Road. Cortez is about 15 miles from the turnoff.

Eureka District

Recorded production for this district is 1,230,000 ounces of gold, all from lode deposits. Large amounts of silver and lead have also come from this district. There were several large mines here, such as the Eureka Consolidated and the Richman Mining Company. These two mines were operated from 1873 to 1906. There have been small amounts of placer gold mined here but no large operations are known. Eureka is located on Highway 50.

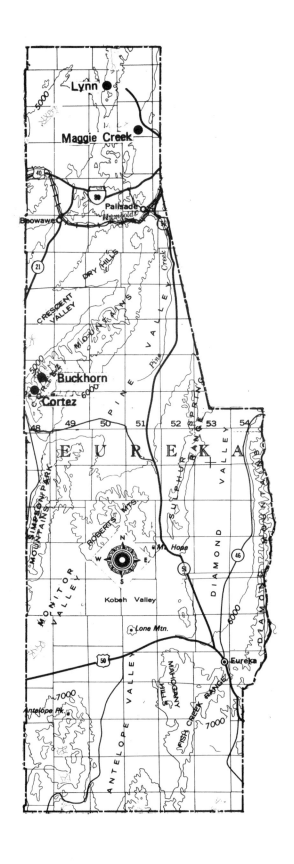

Photo top of hoist house and ore bin. Mine in Eureka County circa 1890.

Lynn District

This was a small placer mining district until the Carlin Mine was discovered in 1962. Prior to the Carlin Mine, total production was set at 9,000 ounces, all placer. In 1979 the Carlin Mine produced 133,600 ounces of gold by itself. The Lynn district is about 18-20 miles northwest of the town of Carlin in the Tuscarora Mountains. The placers were first worked in 1906. This is a large district, and placers have been worked in Lynn, Rodeo, Simon, and Sheep Creeks. The richest gravels are at the upper end of the ravines, and the gold is coarse and angular. The placer gold here is some of the purest in the state. It runs from 920 to 960 fine. Water is a problem here, and by the first of July, the creeks are dry. Lynn Creek was the site of the original discovery and has been very productive.

Maggie Creek District

There has been a small amount of placer gold mined here over the years but there are no records available. This is also known as the Schroeder or

Suzie Creek district. Maggie Creek is north of Highway 80 near the Elko County border line.

Turquoise

There are a couple of spots in the Tuscarora Mountains that have produced turquoise, both in the Lynn mining district. The Number 8 Mine has been the biggest producer, with a million and one-half dollars worth of turquoise mined by the owners. Some large specimens have been found here, one of which weighed nearly 150 pounds and was of excellent color and texture, with good hardness. One 9-pound piece was sold for $1,600. The Number 8 Mine is located on the west slope of the range about 18 miles northeast of Dumphy. The area around Buckskin Mine in the southern part of the county is also said to have some turquoise.

Agate

You can pick up some nice agate in Pinto Canyon. Take Highway 50 southeast to a good gravel road going west. Take this to the entrance of Fish Creek Valley. Look in the mouth of Pinto Canyon.

Palisade Ghost Town

Palisade was once a wild and woolly town, and the junction of three railroads. This is a good location for railroad buffs to use their metal detectors. From Carlin take Highway 51 south 9.6 miles. Then go west 4 miles to Palisade.

Treasure Tale

The only real good treasure story here is shared with White Pine County. Old Jim Pogue was not well-liked, but he didn't care; he had money. He was called the rich miser of Pogue Station. He sold water and feed to passing horses and freight teams as well as to the railroad. He owned the only good well and his prices were stiff. It is said that he made as much as $25,000 a year off his business — darn good money nowadays — but around the turn of the century, when old Jim was in business, that was a fortune. He never put any money in the bank and was not noted as a big spender, so when he died in 1915 people began to wonder where he hid his money. Some local men dug up a few coins in the Station Yard but that's all. The Station is in White Pine County, but his corrals are in Eureka County. A man found a box with $11.00 in it in 1936 but that's all that has ever been located. This would be a good metal detector site. The Station was located on the western side of the Pancake Range of mountains, near Highway 20, just after it crosses into White Pine County.

Humboldt County

This wild, rugged, and beautiful county has a little bit of everything — gold, gems, lost treasure, and ghost towns. It is located in the northern part of the state and borders Oregon. Here you will find the Black Rock Desert, and ghost towns with names like Gouge Eye, Rebel Creek, and Jumbo. The placer deposits were worked here as early as the 1860's, but the largest production has come in this century. Since the area is so remote, and parts of it somewhat isolated, it is worth the effort to prospect here, because your chances of finding underdeveloped and overlooked areas are greater.

Awakening District

This district is located northwest of Winnemucca in the Slumbering Hills near Awakening Peak. Take Highway 95 north out of Winnemucca for about 30 miles to Highway 140. Go west on 140 for 2.3 miles to a road going south. Go 0.8 mile to another road. Follow this road to the road to the ghost town of Daveytown; those are the Slumbering Hills you see. The first recorded production from this district was in 1914. It is estimated that at least 27,500 ounces of gold have come out of these hills. The biggest producer was the Jumbo Mine. The placer deposits were worked in the 1800's but with no great success. The lode deposits are quartz veins in slates.

Dutch Flat District

This is a placer mining area located 18 miles northeast of Winnemucca, on the western slope of the Hot Springs Mountains. The best way to reach the area is to take the good gravel road going north out of Golconda, then take the dirt road that skirts the southern tip of the range to the Dutch Flat area. The gold is found in stream gravels and on the slopes of the hillsides. The richest gravels are on or near bedrock. The gold is both coarse and fine. The coarse gold is rough and angular and some is still in its quartz matrix. There have been a few fair sized nuggets found here. Total production is put at about 10,000 ounces. This would be a good dry-washing area as there is little water available.

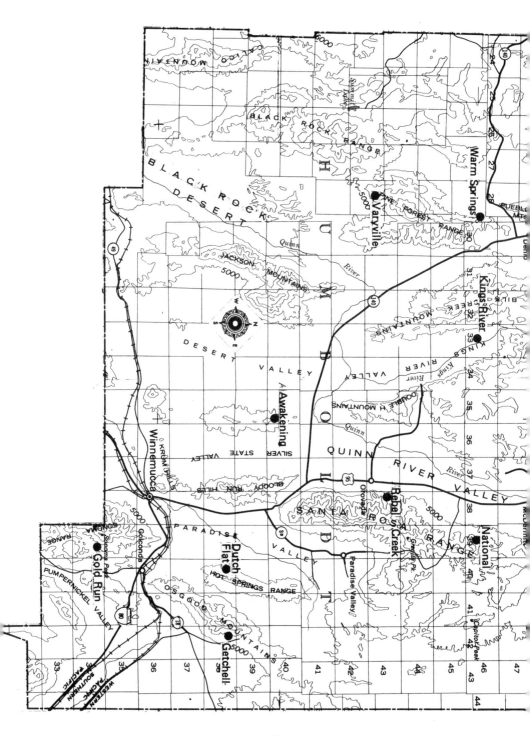

Gold Run District

This district is also known as the Adelaide district. Mainly a lode mining district, there has been some placer gold mined here. Work began here as early as 1866, but the most productive years were from 1878 to 1914. The Adelaide Mine was the principal mine during this period. Production is set at 24,000 lode, and 2,000 placer. There is a good gravel road going south out of Golconda that will take you to this area; it is about 10 miles. Most of the lode gold is a by-product of copper and silver ores.

Kings River District

There are reports of placering being done by Chinese miners in the late 70's along China, or Chinese Creek, and Horse Creek, which drain into the Kings River near the Oregon border.

National District

The deposits here were first discovered in 1907. Some of the ore here was very rich, as high as $30 a pound, in some cases. The first recorded production was in 1909, and the best year was 1911. The district is located on the western slope of the Santa Rosa Mountains, near the Oregon border. To reach the district, take the dirt road 6 miles north of the road to Hinkey Summit, off of Highway 95 south of McDermitt. About 6 miles down this road you come to a fork. Take the road on the left about 2.5 miles to the site of National. The mine is another 1.5 miles farther up. The recorded production for the district is 177,000 ounces. Try your luck in the washes here.

Paradise Valley District

Also known as Mount Rose district, this district is located on the east slope of the Santa Rosa Mountains, 8 miles northwest of the town of Paradise Valley. The deposits were discovered in 1868, and most of the activity took place from 1879 to 1890. Most of the total of 73,400 ounces has been a by-product of the silver lodes, but there have been small amounts of placer gold found here. The town of Paradise Valley can be reached by taking Highway 8B off of Highway 95 north of Winnemucca.

Potosi District

This has been the largest producing district in the county with 486,000 ounces recorded, all from lode deposits. This is also known as the Getchell district and can be reached by taking Highway 18 north off of Highway 80 for about 16 miles until you come to a fork. Take the road going north (18 turns east to Midas) about 10 miles to the Getchell Mine. Some very rich veins have been found here. The first recorded production was in 1938.

Rebel Creek District

This is also known as the Willow Creek or the New Goldfields district. Rebel Creek crosses Highway 95 a few miles north of Orovada. The placer deposits are in the creeks coming out of the Santa Rosa Mountains, all the way from Rebel Creek on the south to Pole Creek on the north, a distance of 17 miles. There was so much gold bearing quartz float on Pole Creek that a 5 stamp mill was set up to work it. The Willow Creek placers in this district were worked in the early days by the Chinese. All of the recorded production took place between 1875 and 1893.

Varyville District

This has been known as the Columbia, the Bartlett Creek, and the Leonard Creek district over the years. By taking the Quinn River Crossing Road west you can reach this location. The road is about 30 miles south of Denio off of Highway 140. The first recorded production took place in 1875. Small scale placer mining has been done here ever since then. Try your luck in Teepee Creek or Snow Creek off of Leonard Creek, and bring your dry washer in the summer.

Warm Springs District

Also known as the Ashdown, Vicksburg, or Pueblo district. Indians burned down the stamp mill here in 1864 and ran off the miners. The Ashdown Mine was the most productive and it was their mill that was burned. The Ashdown Mine was sold for one-half million dollars in the 1930's. The ore is gold bearing quartz veins in granite, and production is set at 24,000 ounces, all from lode. To reach the area take the road going south off of Highway 140 about 9 miles west of Denio Junction.

Winnemucca District

This district is northeast of the town of Winnemucca. The gold and silver veins have a recorded production of 35,000 ounces of gold. There has been a small amount of placer gold found in the washes. The first recorded production was in 1872.

Fire Opal

You can find fire opal in basalt at the Firestone Opal Mine. To reach the mine go north on Highway 95 out of Winnemucca 22 miles to Highway 8B. The Firestone Opal Mine is located 22 miles north of the town of Paradise Valley in the Santa Rosa Mountains.

Opal

A fee is charged at the Virgin Valley Opal Mines, and you can find some gem quality fire opal here as well as common opal. To reach the mines take Highway 140 west from Denio 29 miles to the Virgin Valley turnoff. Go south on this road for 3 miles to where the road forks; the west fork leads to the mines. One of the world's largest fire opals was found here.

Agate

Go west out of Denio about 4 miles on a dirt road and search along the side of the road for agate.

Agatized Wood-Opalized Wood

From Highway 140 take Road 8A for 15 miles to a road on your left going southwest. Three miles down this road and you are in the collecting area. Another 9 miles down this road you can find jasper, chalcedony, and opalized wood on both sides of the road.

National Ghost Town

This is a good location to use your metal detector. Follow directions to the National gold district to reach this site.

Treasure Tales

Humboldt County is the location of one of Nevada's best known treasure stories, The Lost Harden Silver Mine. James A. Harden was a member of a wagon train heading for the California gold fields. The wagon train had camped at what is now known as Double Hot Springs on the edge of the Black Rock Desert, and Harden and another man by the name of John Lambert set out to look for some game. They had no luck in the sparsely covered area, and later Harden stated that they had gone about 3 or 4 miles from the camp site in a northerly direction. While crossing a small ravine a bright glitter in the bed caught Harden's eye. When he picked it up he thought it was lead. The two men gathered up about 30 or 40 pounds of the metal to use for making bullets. Harden stated later that there were several wagon loads of the ore spread all over the ravine. There are conflicting versions of when Harden learned that they were shooting silver bullets, but it was after they had left the area. It was ten years later, in 1859, that Harden and a party of prospectors from Petaluma, California, returned to look for the lode. They were unsuccessful. A year later Harden and another group searched for several months for the lost bonanza. It is thought that Harden went a third time but no records are available to prove this. It is possible for a man to be fooled by the many changing faces of the desert. Flash flooding and sand storms can destroy a landmark overnight. Maybe that's what happened to Harden's Lost Silver Mine.

Another famous treasure tale that is sometimes placed here is the Lost Blue Bucket Mine. Some researchers put it in Oregon. It was in 1845 that a wagon train heading for Oregon passed through the Black Rock country. It was while camped at a spring that some members of the train picked up some of the pretty yellow rocks that were laying around. The kids even filled several blue buckets with the rocks. Most of the stones were lost when several of the wagons overturned while crossing the Deschutes River. It was later that they learned that the yellow rocks were gold. A party of 90 of them tried to return to the area but ran into Indian problems and had to return to Oregon. The golden spring was never found again.

Lander County

At one time the area encompassed by Lander County included what is now White Pine, Elko, and Eureka Counties. Since the county was split and the early records of production incomplete, the total gold production figures vary according to the source. The best estimate would be over 500,000 ounces, both lode and placer. Lander County is the state's most important producer of turquoise, with mines located throughout the county. There are several ghost towns as well as a treasure story or two.

Battle Mountain District

This district is located in the Battle Mountains around the town of Battle Mountain. It is a large district and includes the placer deposits in Copper and Rocky Canyons, Willow Creek, Copper Basin, Cottonwood Creek, Duck Creek, and the area around the old mining camp of Bannock. The lode deposits were first discovered in 1864, but it was not until 1909 that the placers were worked to any degree. Most of the placer mining has been done with dry washers on a small scale. The most productive area has been at the mouth of Copper Canyon and in Black Canyon near Bannock. Most of the gold is coarse and a few large nuggets have been recovered. The richest gravels are found on or near bedrock. The lode deposits are quartz-pyrite veins with native gold. Gold also occurs in copper sulfide veins. Production is set at 150,280 ounces combined lode and placer.

Bullion District

This district is also known as the Tenabo, Campbell, or Lander district. Tenabo can be reached by taking Highway 21 south off of Highway 80 for 30 miles to a dirt road going west. This road will take you the little distance to the camp site. Production for the Bullion district is set at 146,200 ounces lode and 10,400 ounces placer. The district covers an area of 15 square miles with the placer ground being in Triplett and Mill Gulches on the west side of Crescent Valley. The placer gold found here has been very fine and angular. A few small nuggets weighing several pennyweight have been recovered.

Hilltop District

The main source of gold in this district has been from the quartz stringers and masses of pyrite and galena containing silver and gold found on Shoshone Peak. To reach this district go east on Highway 80 from Battle Mountain just a little over a mile. Take the road on your right going south. Follow this road for 15 miles to a road going east. About 1.5 miles down this road is the site of

47

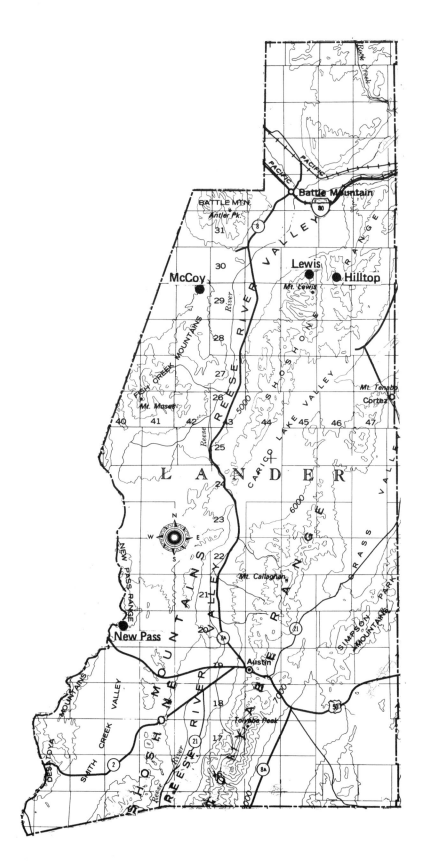

Hilltop. Production is set at 17,400 ounces mostly from lode. Most of the placer gold has come from Crum Canyon. The deposits were first discovered about 1906, and the first recorded production was in 1912.

Iowa Canyon District

This is a small district located about 16 miles north of Austin off of Highway 8A, to the east of the highway. There is a good road into the area. Most of the placer gold found in this district comes from the gravels in the canyon. There is no recorded production for the district, but there was quite a bit of activity here during the 1930's.

McCoy District

This is another small district. There are production figures for only two years, 1929 and 1940. To reach the site of McCoy go south on Highway 8A, 21 miles to a side road heading southeast. McCoy is about 6 miles down this road. The placers are northeast of the town site.

Mud Springs District

This district is also known as the Pittsburg, Dean, and Lewis district. This district is a few miles northwest of the Hilltop district. All of the recorded production is from lode, with the Pittsburg and Morning Star Mines being the largest producers. Placer gold has been mined on a small scale in Mud Springs Gulch, Rosebud Gulch, and Tub Springs Gulch, with the most productive area being Mud Springs Gulch. Total production is set at 51,000 ounces. The source of the placer gold is thought to be quartz veins on Granite Mountain.

New Pass District

This district is west of Austin and north of Highway 50, in the New Pass Mountains near the border with Churchill County. The ore is gold quartz veins in gabbro. Total production is set at 16,000 ounces. Some say that the veins were discovered in the 1860's, but the first recorded production was in 1935.

Reese River District

This is a large silver mining district, with a small amount of gold recorded mainly as a by-product. Total production is thought to be at least 10,000 ounces of gold, from lode. This district is also known as the Austin, Amador, and Yankee Blade district. The location is north of the city of Austin.

Steiner Canyon District

The first gold was found here while digging a well. They also discovered an abundant supply of water with it. Due to the large amount of water, early efforts to mine the placer deposits were unsuccessful. This is also known as the Bobcat district, and is located about half-way between Austin and Battle Mountain, along the Nevada Central Railroad Line.

Home-made dry washer used in Tenabo District, Nev.

Turquoise

Lander County has been the most important producer of turquoise in Nevada. There are mines all over the county, but the two most important districts are Copper Basin and Bullion. To reach the Copper Basin, go south on Highway 8A out of the town of Battle Mountain 4 miles to a dirt road going west. Follow this road about 3.5 miles to Copper Basin. The Turquoise King Mine is 0.5 mile west of Copper Basin and has produced a small amount of gem stones. The largest producer in the district is the Blue Gem Mine which is only about 4,000 feet northeast of Copper Basin. The deposit was discovered in 1934, and quite a bit of work was done here. Some excellent stones have come from this deposit. Production is set at $1,000,000. You can also find some copper minerals here as well.

The Bullion district is on the east side of the Shoshone Range of mountains and is the largest turquoise mining district in the state, and includes the mines around Gold Acres as well as those in the area of Tenabo. The Indians had mined this area for years but it was not until 1938 that the first claim was filed. Some of the most beautiful spider-web turquoise ever mined in Nevada has come from this district. To reach this area go south on Highway 21 from Highway 80 for 30 miles to a dirt road going west; this road leads the little way to Tenabo. Gold Acres is south of here a few miles. There are so many mines and claims here that it is impossible to list them all. Some of the largest producers have been the Steinlich Mine, 1 mile northeast of Gold Acres, with reported production of $500,000; the Super-X or Arrowhead Mine, 2.5 miles northwest of Tenabo, with $200,000; and the Blue Eagle or Blue Sky Mine, about 3.5 miles northwest of Tenabo, with over $1,000,000.

The Cortez or Fox Mine is the largest producing mine in the state and is located about 1.5 miles southwest of the mouth of Cortez Canyon. The town of Cortez is located on Highway 21. Production is set at over 500,000 pounds. The Godber or Dry Creek Mine has produced nearly $500,000 worth of turquoise for its owners. It is located about 5 miles north of Highway 50 near Hickison Summit.

Petrified Wood

Take Highway 8A south from Battle Mountain 10 miles to a fork. Take the road to the right, and follow this road 2 miles to a dirt road. Go 20 miles on this road to a road on your left. Take this road 6 miles to Dacie Creek petrified wood field.

Agate-Opalite-Chalcedony

Go north from Battle Mountain 18 to 20 miles to Rock Creek Road, it is about 15 miles to the collecting area. Search all along the creek for agate, petrified wood, opalite, and chalcedony.

Lewis Ghost Town

Take the good gravel road going south out of Battle Mountain 9.5 miles to road heading southeast. Lewis site is 3.8 miles from the junction. Nearby is the site of the silver mining camp of Betty O'Neal.

Treasure Tale

Somewhere near Hickison Summit is said to be buried a poke of gold buried by a miner from California. According to the story two miners were going home after making their pile in the Mother Lode and decided to camp at the summit overnight. One of the miners buried his gold before going to sleep, as was his habit. The next morning the two men got into a fight and the miner who had buried his gold was killed. His partner got scared and ran away. When he was captured he told the lawmen about the buried gold but they were unable to find it.

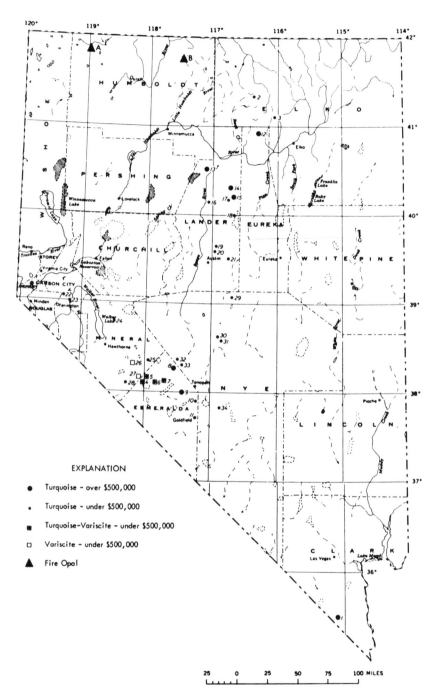

—Gems in Nevada (numbers refer to deposits or districts listed in table 18).

52

CLARK COUNTY

1. Crescent district
   Wood (Aztec, Toltec) mine
   Simmons (Crescent) mine
   Smithson-Phillips mine
   Morgan mine

ELKO COUNTY

2. Rock Creek mining district
   Stampede mine
3. Merrimac mining district
   Carlin Black Matrix mine

ESMERALDA COUNTY

4. Candelaria-Sigmund group
5. Los Angeles Gem Co. group
6. Coaldale mining district
   Holland claim
   Wilson-Capps claim
   Sigmund claim
   Wilson-Capps-Reik-Botts group
7. Carr-Lovejoy group
8. Crow Springs mining district
   Royal Blue mine
   Marguerite claim
   Hidden Treasure (Myers-Bonna) claim
   William Petry mine
   Oscar Wehrend claim
9. Lone Mountain mining district
   Lone Mountain mine
   Livesley mine
10. Klondyke mining district
    Smith Black Matrix mine
11. Goldfield mining district

EUREKA COUNTY

12. Lynn mining district
    Number 8 mine

LANDER COUNTY

13. Battle Mountain mining district
    Blue Gem Lease mine
    Myron Clark mine
14. Bullion mining district
    Gold Acres (Steinich) mine
    Mud Springs mine
    Tenabo group
    Stone Cabin (Steinich) mine

15. Cortez mining district
    Fox Cortez group
16. Pinto Watts mine
17. White Horse mine
18. Jimmey Allen mine
19. Greentree mine
20. McGinnis mine
21. Dry Creek mine

LYON COUNTY

22. Otto Taubert #1
23. Yerington mining district
    Otto Taubert #2

MINERAL COUNTY

24. Rand mining district
25. Pilot Mountains mining district
    Turquoise Bonanza
    Moqui-Aztec mine
    Troy Springs claim
    Copper King claim
    Montezuma mine
26. Silver Star mining district
    Dunwoody-Prichard group
27. Candelaria mining district
    Candelaria mine
28. Basalt mine

NYE COUNTY

29. Indian Blue mine
30. Belmont mining district
    Weber mine
31. Zabrinsky mine
32. Blue Gem (Easter Blue) mine
33. Royston mining district
    Snow Storm claim
    Aztec claim
    C.O.D. claim
    Bunker Hill claim
34. Cactus Peak mine

HUMBOLDT COUNTY

A. Virgin Valley mining district
   Bonanza mine
   Rainbow Ridge mine
B. Firestone Opal mine

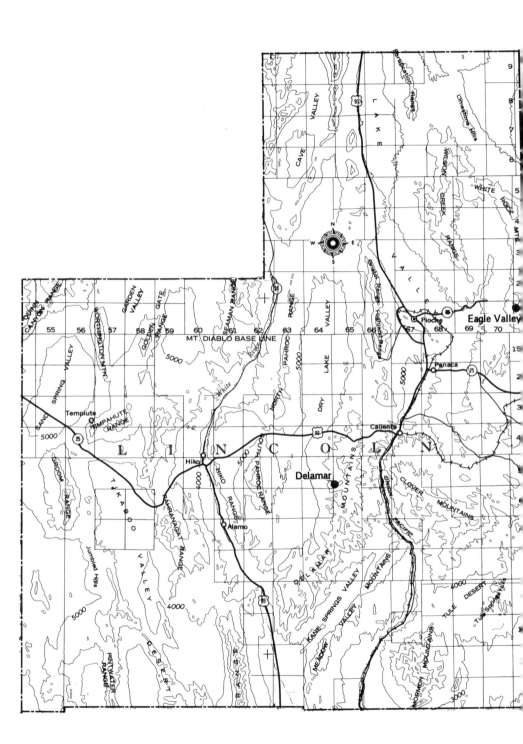

Lincoln County

Even though there are only a couple of gold districts in the county, the Delamar Mine here has been one of the most productive mines in the state. There are a few collecting spots for the rockhound and a treasure tale or two as well as several ghost towns for the metal detector user and treasure hunter.

Delamar District

This district is also known as the Ferguson district. To reach Delamar go west on Highway 93 from Caliente about 17-18 miles to Pole Line Road. Go south on this road about 6 miles to a road going southeast. It is about another 6 miles down this road to the site of Delamar. Most of the production took place here from 1895 to 1909. The first recorded production was in 1892. There was a great deal of activity here again in the late 1930's. Official production is set at 217,200 ounces, all from lode. The miners here were plagued with silicosis from the silica dust in the mines and lost their lives because of it. The deposits are quartzite in breccia containing comb quartz, sulfides, and gold.

Eagle Valley District

There were small amounts of placer gold found here from 1900 to 1904, and then again in the late 1930's. This area can be reached by taking Highway 85 east from Pioche.

Pioche District

More than 100,000 ounces of gold have come from this district, all as a by-product of the lead-silver ores. The mines are located in the area around the present town of Pioche on Highway 93.

Sphalerite

Sphalerite can be found at the Highland Mine and the Ely Mine in Pioche.

Galena

You can pick up some galena at the Monarch Mine in Bristol. To reach this location go north on Highway 93 from Pioche 13 to 14 miles to a good road going west. Follow this road a little over 6 miles to a fork. Stay to your left and go another 4 miles to Bristol. You can also find some samples at the Bristol Mine and the Silverhorn Mine in this area.

Delamar Ghost Town

One of the best sites in the state. Follow directions to Delamar gold district to reach this ghost town. There are several stone buildings still standing and many foundations can still be seen. Good metal detector hunting here. At one time Delamar was the largest city in southern Nevada. Fire destroyed the main part of town and it was rebuilt. When Goldfield and Tonopah boomed, Delamar died.

Railroad Ghosts

By following the railroad tracks south from Caliente you will come across the sites of the many settlements that sprang up, then died. Camp sites of Etna, Boyd, Kyle, Cloud, and Virgo dot the route.

Delamar was a Southern Nevada gold boomtown around 1900. The smoke and dust from the mill created quite a pollution problem in this desert town. Some millers and miners died of silicosis from this "Delamar dust."

Treasure Tales

Delamar is the location of one of the most often told treasure stories of the area. When Delamar was at its peak, they say that one of the officials of the Jackrabbit Mine performed a little highgrading with some help from an assayer. According to the story, they buried nearly $70,000 in gold bullion they had stolen somewhere in town. The mine official died from silicosis before they could smuggle the gold out of town. The mine official had never told the assayer where he was hiding the loot. People still search for the gold as it still has never been recovered.

Somewhere in the Pahranagat Valley about 10 miles south of Hiko $15,000 is said to be buried. The money belonged to two cowboys who camped in the valley while trading horses. They fell out with each other and one of the cowboys killed the other and hid the money. He was caught and hung. Before he died he sent a letter to the dead cowboy's sister confessing his crime and told her he buried the money near where they were camped. The sister made several trips to the area to search for the money but never could find it.

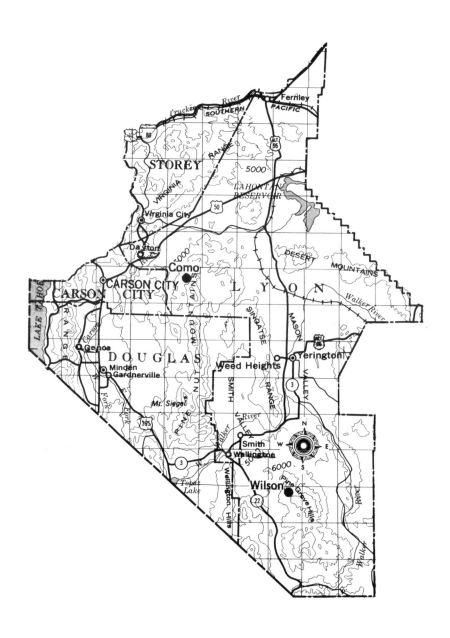

58

Lyon County

It was here that the first recorded discovery of gold was made. This one small county has more gem and gold prospecting sites than some counties twice its size. The gold discovery in Gold Canyon led to the discovery of the Comstock Lode and all the excitement that followed. There is no way to accurately tell how much gold has been mined here. There is evidence that the placer deposits were mined long before the first recorded discovery. One large potential source of gold lies in the tailings along the Carson River. These tailings are from the early-day mills built to treat the Comstock ores. At one time there is said to have been 150 mills in the Gold Canyon district, many of them along the Carson River. These early-day mills employed several different processes of recovery. Compared to present methods, these were very crude processes, and the recovery rate was only 60 to 65 per cent; the other 35 to 40 percent was sluiced down the canyons or deposited directly into the Carson River. Lyon County has agate, jasper, opal, turquoise, and many other gem and mineral deposits for the rockhound. All in all, a must-visit county for the prospector, rockhound, and treasure hunter.

Gold Canyon District

This district is also known as the Silver City, Chinatown, Devil Gate, or Dayton district. It is in Gold Canyon near the town of Dayton which is on Highway 50 about 10 miles east of Carson City. Placer gold was discovered by Abner Blackburn in 1849 in the sands of the Carson River at the mouth of Gold Canyon. Development was slowed by the fact that water is available only a few months out of the year. The placer gold is from the disintegration of the lode deposits of the Comstock and Silver City deposits. Production is set at 143,500 ounces lode and 46,500 ounces placer.

Buckskin District

This district is sometimes placed in Douglas County, and there is no recorded production for the district. It is located in the northern portion of Smith Valley on the border with Douglas County. This is a placer mining district.

Pike Street - Dayton, nev, 1898

The ghost town of Dayton, 12 miles northeast of Carson City and on the north bank of the Carson River, was a firmly established community based on silver milling, farming and teaming. The hundreds who lived here are gone, and except for the paving of the street, the scene looks much the same today.

Como District

This is a small lode mining district located about 10 to 12 miles southeast of Dayton. There is a road leading into the district coming out of Dayton but it is rough and should only be traveled in the summertime. The deposits were first worked around 1864 and production is recorded as starting in 1900. Production is estimated at 10,000 to 15,000 ounces, all lode.

Assay Office in Como District, Lyon County, circa 1902.

At Silver City, just four miles south of Virginia City and the Comstock Lode, there was quite an ethnic community, as evidenced by the names of these saloons in the business district.

Wilson District

Recorded production for this district is 408,000 ounces, all from lode. This district is also known as the Pine Grove, or Rockland district. It can be reached by going south on Highway 3C from Yerington. About 2.5 to 3 miles after the junction with 3C and after you cross the east Walker River, you come to a road going south. Take this road about 10 miles to a road heading west. A mile down this road you will come to a fork. The left fork leads to the site of Rockland; the right fork leads to the site of Pine Grove. The ore is quartz veins containing gold and pyrite in quartz monzonite.

Yerington District

This placer mining district is also known as the Ludwig or Mason district. The copper deposits were worked here as early as 1864, but the placer gold deposits were not discovered until 1931. Most of the activity has been in Big Canyon, with minor work being done at Lincoln Flat about 2 miles southwest of Big Canyon. A few nuggets weighing as much as one-half ounce have been found in Big Canyon. There are no records on the production for this district. The source of the gold is thought to be veins at the head of the canyon and an ancient Tertiary river channel. The district is located about 8 miles northwest of the town of Yerington on the west side of Mason Valley between Mason Pass and Gallagher Pass.

Turquoise

The best collecting area today is 8 miles north of Yerington along the Walker River, then northwest 2 miles to Walker River Valley. The largest producers have been the Taubert Mines. The Taubert Number 1 is located about 7 miles west-northwest of Yerington, and 1 mile west-southwest of Mason Pass. The turquoise mined here ranges from dark blue to bluish green to green, much of which is very hard. The Taubert Number 2 is in a group of small hills about 1.5 miles north-northwest of Yerington. The turquoise found here is soft, pale blue turquoise, and some darker, pure blue material.

Opal-Petrified Wood

To reach this collecting area take Highway 95A southeast from Fernley 5 miles to a road going west. The material is spread along the side of this road for 7 or 8 miles.

Agate-Jasper

Go south from Fernley on 95A for 5 miles to a road going west. Follow this road about 2 miles to the collecting area.

Chalcedony-Agate

Go 7-8 miles south on Highway 95A from Fernley. Prospect on the east side of the highway all along here.

Calcite-Gypsum-Iceland Spar

Go south 5 miles past Weeks on Highway 95A to a road heading northeast. One-and-a-half miles down this road is the collecting area.

To accommodate everyone working on the Comstock Lode, smaller mining camps were built around Virginia City, including this one in American Flat. At left is the residential area, while on the right are the recreation hall, dining hall, hospital and store.

Petrified Wood-Jasper

Go 10 miles east of Yerington on Highway 95A. Search on the south side of the highway.

Fort Churchill Ghost Camp

This is now a state park and there is no treasure hunting, but it is worth seeing if you are in the area. You can reach Fort Churchill by going south on Highway 95A; it is about 8 miles past Silversprings.

Treasure Tales

There is said to be a Chinaman's poke of gold hidden somewhere in Gold Canyon. Using your metal detector in any old mining district is always worthwhile. You have to remember there were no banks in those early days and the only place to hide your valuables was in the ground.

Mineral County

Up until February of 1911, Mineral County didn't exist. It was in that year that the county was formed from part of Esmeralda County. Another interesting sidelight concerns the mining camp of Aurora. When the town was first formed in 1860 they didn't know if the location was in the state of Nevada or California, and the town was governed by both states until 1863. Until an official government survey was made in that year, Aurora served as county seat for two states and voted twice in all elections. The site is 3 miles inside Nevada. Little remains of this one-time town of 10,000 today. There are some good collecting areas for the rockhound here, and one of the best known treasure tales is also based in this county.

Aurora District

To reach this district from Hawthorne, go south on Highway 31 for 4 miles to a side road heading west. Take this road about 18 miles to Fletcher Junction. Take the road to Bodie from here about 1.5 miles to a side road heading southwest. Aurora is about 4.5 miles down this road. Since there was so much confusion in its early history, the production figures are only estimated. The minimum is set at 93,600 ounces of gold, all from lode. The ore is gold and silver bearing quartz veins in biotite-quartz latite and andesite.

Bell District

This lode mining district is located in the Cedar Mountains near the border of Nye and Esmeralda Counties. The gold-bearing quartz veins in rhyolite have produced about 34,000 ounces of gold. There have been reports of some placer gold being found here, but there is no recorded placer production.

Candelaria District

This is the second richest district in mineral production in the county. (The Aurora district is the most productive.) The silver and gold bearing sulfides were first discovered in the 1860's, and the deposits have been worked off and on in various degrees since. The recorded gold production is set at a minimum of 13,000 ounces of gold, all as a by-product of the rich silver ores. The early production may have been much greater, but there are no records available. To reach this area go south on Highway 95 from Mina. Ten miles past the junction with Highway 10 you will come to a turnoff heading west. The site of Candelaria is about 7 miles down this road.

Garfield District

Estimates of the gold produced from this district range from 10,000 to 50,000 ounces, all from silver-gold ores. The ore is quartz veins in volcanic rocks, and in limestone. This district is in the mountains northwest of Mina.

Gold Range District

This district has a recorded production of 97,000 ounces of gold, all from lode. The ore here branches out from two major faults and is found in Tertiary rhyolite and in some sedimentary rocks. This district is in the Excelsior Mountains near Mina.

Hawthorne District

The Hawthorne district is located around the town of Hawthorne. The placer deposits are found at Laphan Meadows, near Mount Grant, and in Baldwin Canyon several miles south of Laphan Meadows. The recorded production is only 10,000 ounces of gold, all as a by-product of the rich silver ores. Some nuggets weighing an ounce or so have been found here according to reports.

Mount Grant District

There have been reports of placer deposits being worked on the west slope of Mount Grant, west of the city of Hawthorne; but no records are available as to the amount of gold recovered.

Mount Montgomery District

This district is combined with the Oneota district in most reports. The ore is quartz veins containing gold and silver found in Tertiary volcanic rocks. Total recorded production is at least 10,000 ounces and would be much greater if all production had been recorded. This district is located in the White Mountains southwest of Basalt near the junction of Highway 10 and Highway 6.

Pine Grove District

This district is also known as the Wilson or Rockland district and is shared with Lyon County. See Lyon County.

Rawhide District

The recorded production for this district is 51,000 ounces from lode and only 2,000 from placer. Old timers will tell you that the total from the placer deposits is at least 10,000 ounces. The lode deposits were discovered in 1906 and the placer a short time later. Most of the placer gold has been recovered by small scale dry-washing methods, and as many as 100 men have been known to work here at one time. The best placer ground is an area about one-half mile wide on the southwest slope of Hooligan Hill and in Rawhide Wash. The gravels are angular and a few nuggets weighing as much as two ounces have been found here. The lode deposits are in a network of veinlets in rhyolite, dacite, and andesite. The easiest way to reach the Rawhide area is to go south on Highway 31 from Highway 50 about 20 miles to a fork. Take the road on your right; follow it just about 3 miles to a junction with a road running east and west. Go west on this road 1.5 miles to the site of Rawhide.

Nevada's star mining camp of 1908 was Rawhide, shown here near the height of the boom. A fire destroyed most of these buildings in August 1908, and those which survived were destroyed by vandals and the weather.

Telephone Canyon District

The Telephone Canyon district is located on the west side of the Pilot Mountains, in southwestern Mineral County, 5 miles east of Sodaville. Placer gold was found near the mouth of the canyon during the depression, but there is no recorded production.

Turquoise

The east side of the Pilot Mountains, about 5 miles northeast of Pilot Peak, is the location of the Turquoise Bonanza Mine deposits. The turquoise occurs here as cementing material, and veinlets in an altered brecciated zone, with quartzite. The turquoise found here was an excellent blue color, but some showed a slight green tinge. Production is set at about $400,000.

There are two mines located at the south end of the Pilot Mountains, about 14 miles southeast of Mina. The first is the Moqui-Aztec or S. Simmons Mine. Some excellent light-blue stones with fine brown markings have come from here. The Montezuma or Troy Springs Mine produced some good stones, but the majority of the material was low grade.

There are a couple of mines located about 4 or 5 miles north of Basalt. The Blue Jay Gem Mine is on the east side of Highway 10, and the Blue Gem Mine is on the west side of the highway. Small amounts of turquoise have come from these locations.

Jasper

At Sodaville on Highway 95 take the road going southwest about 1.5 miles to the collecting area. Search on both sides of the road.

Geodes

Go west from the junction of Highway 6 and Highway 10 at Basalt, about 7 miles to a side road going south. About 1 mile down this road is the collecting area.

Pamlico Mill. Owned by governor of Mineral County.

Agate-Jasper-Limb Wood

A little east of Basalt on Highway 6 is a side road going south. Take this road 1.5 miles to a fork. Take the left fork 2 miles to the collecting area.

Obsidian

Go west 11 miles from the junction of Highway 10 with Highway 6 at Basalt to a side road going north. Take this to a trail heading northeast. The collecting area is about 1.5 miles up this trail.

Agate/Crystal Geodes

From Luning go north on Highway 23 about 37.5 miles to a side road going west. Follow this road about 3 miles to a fluorspar mine. The specimens are spread over an area of several miles at this site.

Belleville Ghost Town

About 17 or 18 miles north of Basalt on Highway 10 is the ghost town of Belleville. The site is marked and easy to locate. The town was formed around 1873 and at one time there were at least 600 residents. It had a large business section with hotels, saloons, and so on. This is a good place to use your metal detector and your camera.

Treasure Tales

Around 1908, a prospector named Cody was working in the area around Cedar Mountain. He left his camp one day, heading for Golddyke, and somehow got lost. According to the tale he wandered for three days before reaching Golddyke. Somewhere in his wandering he climbed a small hill trying to sight some sort of landmark. He claimed that he was able to see Paradise Peak to the north from the hill. At the top of the hill he found an iron stained outcropping of white quartz. He took several samples from the ledge with him and set out for Golddyke. The ore was a high grade deposit he learned after having it assayed. He was never able to find that hill again, and until he died in 1930, he never stopped searching for the ledge.

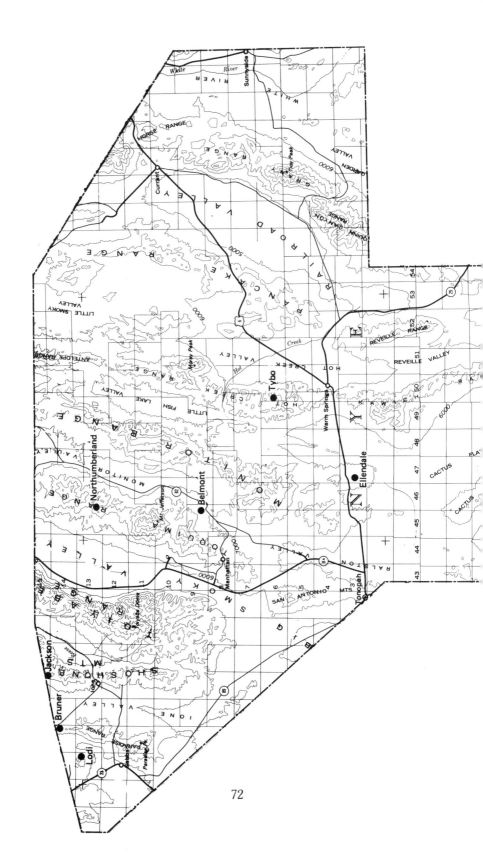

72

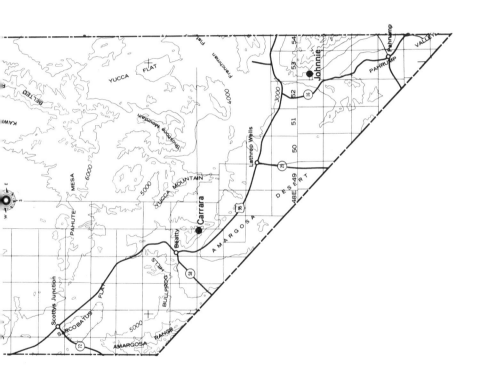

Back in 1905 Beatty, adjacent to Death Valley, was a mining camp of promise. Tents, wood frame businesses and stables were all thrown together without much regard to planning or health standards.

Nye County

Nye County is one of my favorite counties in which to go prospecting or treasure hunting. It is one of the largest counties in Nevada, and has more gold districts than any other county. The ruins at Rhyolite are probably the most photographed of any ghost town. The dumps around the town of Tonopah have long been a favorite of bottle collectors. It can get cold up in the northern part of the county during the winter, and the wind can cut like a knife if you are not protected. There's a big celebration every year in Tonopah that you might like to see; ask one of the locals there about it. There are some good collecting areas for the rockhound and some of the best metal detecting locations you can find anywhere. Try to spend some time here. You'll be glad you did.

Beatty District

During the early 1920's there was a small rush to the Amargosa River on both sides of the town of Beatty. Quite a few claims were filed, but there is no recorded production for the district. A little gold was found, but not enough to be of any economic importance, and the miners moved on to greener pastures.

Bruner District

The gold and silver quartz veins in Tertiary volcanic rock have produced about 17,200 ounces of gold here. The principal mine of the district was the Penelas. There are reports of small scale mining here as early as 1905, but the bulk of the production was in the 1920's and 1930's. This district is located near the border with Lander County in the northwest portion of the county.

74

Bullfrog District

This district is the perfect example of boom and bust gold mining. The first gold was discovered in 1905, and in two years it was a roaring camp of 12,000 souls. They built modern stone buildings to house its many stores. There were banks, newspapers, gambling halls, countless saloons, an ópera house, and a large Union Station. Three railroads had lines built into Rhyolite to carry all the people to the new bonanza, and to carry out the gold. By 1911 the ores had already begun to play out. Today the empty streets are lined with crumbling walls and the foundations of the long gone buildings. Only the Union Station and a few other structures, such as the Bottle House, are occupied now. The recorded production for the district is 120,400 ounces. Quite a bit of silver was also mined as a by-product. The pay ore was crustified gold-bearing quartz veins in Tertiary rhyolite. To reach this location go west out of Beatty on Highway 58 for 4 miles to the Rhyolite turnoff; it will be on your right.

Carrara District

This district is located about 8 miles south of Beatty at the western slope of Bare Mountain. Several claims have been filed here, but there is no recorded production. Some gold washed down from the veins in limestone that are known on Bare Mountain, but not enough to be mined on a large scale.

Cloverdale District

This district is also known as the Golden or Republic district. It is located about 40 miles northwest of Tonopah in Cloverdale Canyon, along Cloverdale Creek. The placer deposits were first discovered in 1906, and small scale placer mining has taken place here since then. Due to the lack of water most of the year, all the work has been done by dry washing. There is no recorded production for the district.

Ellendale District

This is a small district east of Tonopah and south of Highway 6 in the hills west of Stone Cabin Valley. The ore is in irregular veins filled with gold-bearing iron stained quartz in rhyolite near its contact with andesite porphyry. Production is set at roughly 8,000 to 20,000 ounces, all from lode. The first recorded production was in 1908. There has been little work done here since the late 1930's.

Gold Hill District

This district is located in the southern end of the Toquima Mountains, about 6 miles north of Round Mountain. The ore is gold-bearing quartz veins in rhyolite. Production is set at about 25,000 ounces, all from lode. This area is now part of the Toiyabe National Forest.

Jackson District

This district is located on the western slope of the Shoshone Mountains near the Lander County line. The ore is quartz-sulfide veins in metaandesite. Production is set at between 24,000 and 48,000 ounces, all from lode. To reach this area take State Highway 91 east off of Highway 23 about 2 miles north of Gabbs and go about 16 miles to a road going north to the site of Ione. These are the Shoshone Mountains.

Jefferson Canyon District

This district is in Jefferson Canyon just east of Round Mountain. This is also a rich silver district. The first recorded production was in 1869, and the mines were active until 1890. There was activity again in the late 1930's and some production was recorded. The ore is a sulfide vein along contact of Ordovician limestone and Tertiary porphyry. The production is estimated at 20,000 to 25,000 ounces, all from lode.

Johnnie District

Not much remains of this camp today. It is located on Highway 16 about 11-12 miles southwest of the junction with Highway 95. The deposits are on the east slope of the Montgomery Mountains. The lode deposits were discovered in 1903, and the placer deposits in 1921. Some claim that the Mormons had worked the placers in the early days but no records are available in regard to that period. The placers run for a distance of 10 miles, starting about 2 miles north of the Johnnie Mine and running to about 4 miles south of the old Congress Mine. Some placer gold can also be found on the western slope of the Montgomery Mountains. Water is scarce and the placers have to be worked by dry-washing. The ore is gold-bearing quartz veins in quartzite, shale, limestone, and a conglomerate. Production is set at 40,000 ounces lode, and at about 20,000 from placer.

Lodi District

This district is also known as the Mammoth, Marble, or Ellsworth district. Production is recorded as early as 1866 and has continued off and on since that time. Production is set at from 10,000 to 20,000 ounces, most all as a by-product of lead-silver ores. There was a 10-stamp mill at Ellsworth that served all the district. Ellsworth can be reached by taking Highway 91 east off of Highway 23, about 2 miles north of Gabbs, and going 16 miles to a side road heading north. Follow this road 7 miles to a crossroad. Turn left and go about 3.5 miles to another side road going southwest. Ellsworth is about 6.5 miles down this road.

Manhattan District

This is one district that you should visit if you are interested in Nevada gold mining. Mining was recorded here as early as 1868 and is still going on today. The Summa Corp. spent millions here recently in an effort to rework the tailings from the early day operations. When I was here last, there were a great many pieces of heavy equipment just rusting away. Some of the equipment cost over $100,000 each. There were a couple of problems; the first was Howard Hughes' death and the tying up of the Summa Corp.'s assets, and a lack of practical knowledge on how to work the tailings. A nice young couple have taken over the old water and power plant and make their home and business there. He has a small mine he operates, and she has a rock shop. He milled me a little of his ore and it was quite rich. Stop and see them for information on the various areas and claims. Manhattan can be reached by going east on Highway 6 from Tonopah 5 miles to Highway 8A. Go north on this road about 37 miles to State 69. Go east on this road 7 miles to Manhattan. The altitude is 7,000 feet and it is cold in the winter months so be prepared.

The placers were discovered in 1905 when one of the lode mining companies was digging a well and colors were found in the dirt wall of the well. The discovery caused little excitement at the time. In 1906 a miner found some nuggets on the surface and efforts to mine the placers began. Most of the placer gold has been recovered by drift mining the gravels, but a lot of the early production came from dry-washing the surface deposits. There are three types of placer deposits here, the older bench gravels, the buried gravels of the gulch, and the recent wash deposits on the hillsides and in the dry stream courses. The gold is both fine and coarse and is found throughout the gulch. During the early days finding nuggets weighing up to an ounce was not unusual. The lode deposits are in Cambrian limestone and quartzose schist in a hanging wall of thrust. Production is set at 280,000 ounces lode and 206,000 placer.

Round Mountain District

Both the lode and placer deposits were found in 1906. It is located about 15 miles north of Manhattan on the east side of the Toquima Mountains. To reach this district go east on Highway 6 for 5 miles from Tonopah to Highway 8A. Go north on Highway 8A about 48 miles to Highway 92. Round Mountain is 3 miles down this road. Recorded production is about 390,000 ounces lode and 150,000 ounces placer. The main placer area is on the south and west slopes of Round Mountain. It is of the residual type of deposit and none of the gold has traveled more than a few hundred feet from its source. The gravels range from a few feet to fifty feet in depth and are cemented near bedrock by a limey deposit. Most all the nuggets found are encrusted with quartz or siliceous limonite. The first placer mining done was with dry-washing methods, but large scale hydraulic mining began as early as 1907. Large scale placer mining was carried on as late at 1959. The lode deposits are complex gold-bearing fissure veins of Tertiary age.

Millet District

There are reports of small amounts of gold being found in Clear Creek and Ophir Canyon in the Toiyabe Mountains about 75 miles north of Tonopah. There is no recorded production for this district.

Northumberland District

Most of the recorded production of 35,400 ounces lode came from open pit mining in the late 1930's and early 1940's. This is mainly a silver district, with the gold as a by-product. The first recorded production was in 1866 and continued steadily until the early 1890's. To reach this district go north on State 82 off Highway 8A about 53 miles to a road going northwest. Six miles down this road is Northumberland Cave. The site of Northumberland is another mile up the road from the cave.

Tonopah District

Shortly after its discovery in 1900 by James Butler, the Tonopah district became the most productive district in the county. The peak years were from 1910 to 1914. The deposits are in the San Antonio Mountains which tower over the town. The ore is replacement veins in faults in Tertiary volcanic rocks. Even though it is considered primarily as a silver district, it has a recorded production of 1,880,000 ounces of gold. Tonopah is on Highway 6 and a good place to pick up supplies, as it is still a good-sized little town.

Tybo District

The deposits in this district are said to have been discovered when an Indian led an old prospector to a rich outcropping of lead-silver ore. For a period before 1900 this was the most productive district in the county. Tybo is in the Hot Creek Mountains northeast of Tonopah. To reach this area go east on Highway 6 from Warm Springs about 9 miles to a side road on your left heading north. Follow this road about 5 miles to a fork. Take the left fork about 4 miles to the site of Tybo. The recorded production is 27,300 ounces of gold, all as a by-product of lead-silver ores.

Union District

This district is also known as the Berlin or Ione district, and is located on the west slope of the Shoshone Range of mountains. Most all of the recorded production of 10,000 ounces was recovered prior to 1900. The district was discovered in 1863 and the amount of placer gold recovered has been small. People still find a little gold today using drywashers. To reach Ione go north on Highway 23, off Highway 95 about 34 miles to Highway 91. Follow this

80

road about 16 miles to a junction; take the side road on your left going north. Stay on this road about 7 miles to another junction; turn right on this road. Ione is just a mile or so down this road.

Turquoise

The Royston district, which is located on the Nye County border with Esmeralda County, about 24 miles northwest of Tonopah, is one of the better known turquoise producing areas in Nevada. The mines are scattered over an area nearly a mile long in a shallow canyon. The Royal Blue is the main mine of the district. The mine was discovered in 1906 by two prospectors named Workman and Davis. The turquoise ranges in color from dark sky blue to a pale light blue. The Royal Blue Mine has been one of the major producers in the state. At one time the mine produced as much as 1,250 pounds of turquoise a month. Turquoise has been found on the eastern slope of the Toquima Mountains about 1.5 miles south of the border with Lander County. Small amounts have also been mined around Belmont and Manhattan.

Opalized Wood

There is a large collecting area northwest of Gabbs. It can be reached by going north on Highway 23 about 1 mile to a side road going west. Take this road about 9 miles to a road going north. Follow this road about 2 miles. Collecting starts when you leave Highway 23.

Geodes-Fluorescent Chalcedony

To reach this collecting area go north on Highway 95 for 4 miles to a side road on your right heading northeast. Follow this road for about 5 miles to the collecting area.

Amethyst

You can find some amethyst in the Bullfrog mining district around the ghost town of Rhyolite. Ask at the Bottle House for best collecting area currently. To reach this area go west on Highway 58 from Beatty 2 miles to the Rhyolite turn-off; it's about 2 miles down this road.

Marble

To reach this collecting area go south on Highway 95 from Beatty about 8 miles to Carrara. Take the side road going east for about 3 miles to the quarry.

Obsidian-Jasper

Try along the west side of Highway 95 about 4 to 5 miles north of Beatty.

Belmont Ghost Town

Most people think of Rhyolite when you mention Nye County ghost towns. You should see Rhyolite; it is worthwhile. You should also see Belmont. Belmont is a little more remote and has not had as many visitors. Belmont was settled in 1865 and had a population of 10,000 at one time—by 1903 it was a ghost town. It was the County Seat for Nye County for a while and the old courthouse ruins are still standing as well as other remains. To reach Belmont take Highway 8A north off Highway 6 about 5 miles east of Tonopah. Go north for 13 miles to the junction with Highway 82. Belmont is about 28 miles up this road.

Treasure Tales

There are several tales of lost treasure in Nye County. One of these took place in the area of Manhattan. One of the early residents of Manhattan was a miner named Sam. According to the story, Sam had done pretty well working the placers in the gulch and had a gallon jar about half full of gold nuggets and dust. Like most of the old timers, he didn't trust banks and kept his poke buried. He owned three houses on the edge of town and most folks think he had the jar buried right there on his property. He told a few friends that he could sit on his front porch and see where his gold was buried.

Another area that spawned several treasure tales is around the old mining camp of Tybo. One story deals with the man who owned and operated the charcoal kilns. He is known to have received $5,000 for a large order of charcoal a few days before he was killed in an accident. He, too, was never known to put his money in the bank. Most people figure his money is buried somewhere near the old charcoal kilns.

Carson City Area
(formerly Ormsby County)

There is no recorded gold production for what was formerly known as Ormsby County. The mills that operated along the Carson River during the 1860's, 1870's, and early 1880's got their ore from the Comstock Lode. There are scattered reports that small scale mining was done here. Whatever was done must not have been too rewarding.

Opalized Wood-Agate

To reach the collecting area take Highway 50 east for 4 miles from Carson City to the Brunswick Canyon turnoff. Follow this road along the river to the bridge leading into the canyon. Collecting is done all along the canyon on both the sides and along the floor.

Epidote

You can find epidote in Vicee Canyon northwest of Carson City. Go north on Highway 395 to Winnie Lane. Go west on Winnie Lane to a road leading up Vicee Canyon. The deposits are in decomposed granite.

Treasure Tale

This story takes place in the 1880's according to some tellers. A stage-coach left Virginia City with a load of gold bullion heading for the mint in Carson City. Four bandits held up the stage outside of Carson City and got away with the gold. Either the guards or a posse were able to kill some of the bandits before they could get too far. The other robbers were forced to bury the heavy loot somewhere in the hills northeast of Carson City. The last of the hold-up men was captured and sent to the Nevada State Prison which was on the outskirts of Carson City. They figured the bandits had not been able to recover the loot before they were captured or killed. The prisoner refused to tell where the gold was buried. The Wells Fargo Company got the bandit pardoned after eight years in the hopes that he would lead them to the gold. He had contracted T.B. in prison and was never strong enough to return to the spot. There are some who claim that the warden of the prison in 1935 considered having the prisoners search for the gold but there are no records of this ever being done.

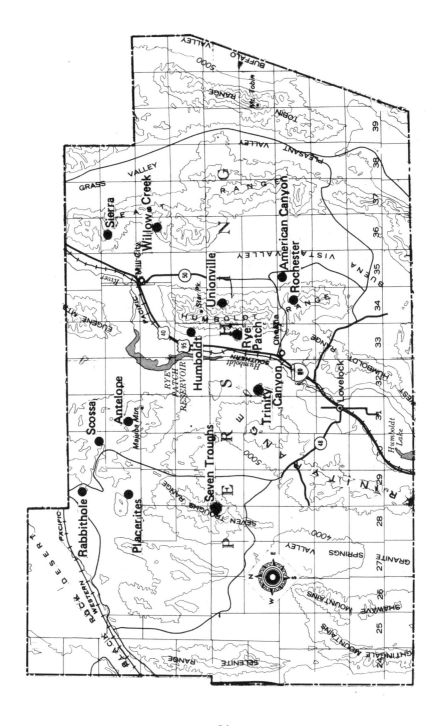

Pershing County

Up until 1919, what is now Pershing County was part of Humboldt County. The Chinese miners were quite active here and because they were so secretive about the amount of gold they recovered, and because so little of it was ever seen at the U.S. Mint, the recorded production figures for the different districts in the county are probably lower than the actual figure. Some of the best drywashing areas in the state are located here, as well as many good gem collecting sites. For the treasure hunter there are several ghost towns to check out and a treasure tale or two they might want to investigate.

Antelope District

There is no recorded production for this district, but there are reports of small amounts of placer gold being recovered on the eastern slope of the Antelope Mountains near Najuba Hill. There was some recorded production from the silver-lead deposits with by-product gold starting in 1912 and continuing until 1928. The placers were first worked during the depression years.

Humbolt District

This district is also known as the Imlay or Eldorado district. It is located in the northern part of the Humboldt Mountains. To reach this district go north on Highway 80 from Oreana to Imlay. A few miles south of Imlay is a road going east into the mountains. This road will take you to Humboldt City; it is about 3.5 to 4 miles. The lode deposits here are in a gold and silver bearing quartz vein in shale, quartzite, and limestone. Some placer mining has been done in Imlay and Antelope Canyons. The recorded production is all from lode and set at 35,000 ounces.

Placerites District

This is another small placer gold district in the northern portion of the county. The gold is found in low lying gravel hills near Rabbit Hole Creek. The placers were first located by a prospector named Mahogany Jack and his three partners in the 1870's. There was a large amount of activity here in recent times, but there is no record of how much gold was recovered. There are at least five gulches in the district that are known to have gold. Cloudbursts have obscured the early diggings, and it is hard to judge how extensive they were. To reach this district take Highway 48 west out of Lovelock 14 miles to a junction. Take the right fork going north; it is about 22 miles up this road to Placerites. Stay to the right to go to Placerites, or you can take the left fork and go 7 miles to the site of Vernon and follow that road to where it rejoins the road to Placerites.

Rabbit Hole District

This is also known as the Rosebud district. The placers here were known as early as 1900, but no real effort was made to work them until 1916. The Rabbit Hole district is north and west of Rabbit Hole Springs and is about 5 miles northeast of Placerites. In the early days small nuggets could be picked up in Coarse Gold Canyon after a cloudburst, according to some oldtimers. There are no recorded production figures for the district. In 1935 this was one of the most important placer mining districts in the state. In that year there were at least 150 men working the deposits with drywashers and rockers. The most productive areas are at the heads of the ravines and on the slopes above the ravines. This is still a popular area for prospectors.

Rochester District

The Rochester district is in the Humbolt Mountains northeast of Lovelock. The placers were first located by a group of miners from Rochester, New York, in the 1860's. These first operations did not create a great stir, but the discovery of rich silver veins in 1911 started a small boom. The recorded production for the district is 70,000 lode and the estimated production for the placers is listed by the state at from 4,500 to 50,000, which is a huge difference. According to some private estimates the placers may have produced 500,000 ounces, but there are no records to support these claims. The early placer workings in Rochester Canyon have been covered by large amounts of cyanide mill tailings. The gravels are from 50 to 100 feet thick and carry values at all levels. The placer gold is coarse and crystalline and runs about 800 fine. Some of the other canyons that have been productive in the past are Limerick, which is about a mile north of Rochester, Walker and American, which are on the eastern slope of the range, and Sacramento and Weaver Canyons. Sacramento Canyon is sometimes listed as a district itself. To reach Rochester go north on Highway 80 from Lovelock about 14 miles to Oreana. Take the road going east for about 3 miles to a side road going south. Take this side road about 4 miles to a fork. Take the left fork. The site of Lower Rochester is about 1 mile down this road and Rochester is another 3 miles past Lower Rochester on this road.

Mazuma, in Pershing County, experienced the unfortunate occurrence of a flash flood which wiped out all of the buildings shown here, except for the mill at upper left. Ruins and debris of the flood still exist below the mouth of Seven Troughs Canyon.

Rye Patch District

This is a small district located about 4 miles north of the Rochester district. Production is set at about 10,000 ounces, all from lode. The ore bodies are irregular fissures in Triassic limestone filled with brecciated wall, quartz, calcite, and ore minerals. The principal mining done here has been for silver.

Sawtooth District

This district is shared with Humbolt County and is placed there by some people. The placer deposits were discovered in 1931. An unusual feature of the placers is that the most productive of the gravels were found at a shallow depth. Much of the gold has been in gravels laying on a layer of a false clay bedrock and the gravels are only 8 inches to 2 feet deep. Try any area where you find white quartz on the surface, as these are usually the places where you will find the gold. Some fair-sized nuggets have been found here. This is a good drywashing area. This district is located about 12 miles north of the old mining camp of Scossa, in the area around Sawtooth Knob.

Seven Troughs District

All the recorded production from this district is from lode, but there are reports of small amounts of placer gold being found here over the years. The deposits were first worked in 1908 and production has continued up until today in various degrees. Total production is set at 160,200 ounces. The ore occurs in mineralized breccia zones and fissures in Tertiary volcanic rocks.

Other minerals found here are copper, silver and lead.

To reach Seven Troughs take Highway 48 west out of Lovelock 14 miles to the fork. Take the right hand fork; this is a good gravel road. Follow this road about 5 miles to another fork; take the left hand fork to Seven Troughs. It is about 10 miles.

Sierra District

Some rich placers were found in this district. It is also known as the Dun Glen, Oro Fino, Sunshine, or Chaffee district. The first lode discovery was made in 1863, and the placer deposits were found shortly after that. The lodes occur as gold-bearing quartz-sulfide veins that are associated with diabase dikes in volcanic rocks. The placer deposits are in gravels and occur over a large area. The most productive placers have been in Auburn, Barber, Wright, Rock Hill, Dun Glen, and Spaulding Canyons. Production is set at 194,000 ounces lode, and 47,000 ounces placer. The Chinese placer miners were quite active here in the 1880's and 1890's. An attempt was made to mine the placers in Barber Canyon in the early 1930's. To reach this area go north on Highway 80 for 1 mile from Mill City to a side road going east. Follow this road to the site of Dun Glen; it is about 9 miles.

South American Canyon

This is a small district that is located about 1 mile south of American Canyon in the Rochester district. The canyon was worked quite extensively by Chinese miners using rockers and pans for many years. There are no records as to the amount of gold mined, but judging from the amount of work done they must have recovered a large amount of gold. About half-way up the canyon there is a bowl-shaped depression that was a natural trap for the placer gold. This depression is marked with shafts and tailing piles as evidence of the Chinese miners' efforts. Most of the shafts are under 15 feet deep. As in most areas worked by the Chinese the best areas have been pretty well worked out.

Spring Valley District

This is a rich placer mining district on the east flank of the Humboldt Range about 5 miles east of the Rochester district. The lode deposits were discovered as early as 1868, but it was not until 1881 that the rich placers were discovered. By the middle 1930's the Spring Valley and American Canyon placer areas had attained the largest output of placer gold in the state. This is another area where the hard-working and thorough Chinese miners were active. In the years between 1884 and 1895 there were an estimated 3,000 Chinese working the placers here. Production is set at 534,000 ounces, all from placer. The most productive areas are in American, Limerick, Spring

Valley, Troy Canyons and Dry Gulch. A dredge was brought in in 1910 by the Federal Mining Company to work the gravels in the lower part of the canyon. They were hampered by the fact that water had to be piped in from Indian Creek which is 5 miles away. The richest placers are found on a layer of false bedrock of clay. The gold is coarse and has a fineness of about 700. Some small nuggets have been found here.

Chinese placer miner with rocker and other tools.

Trinity Canyon

There are reports of small amounts of placer gold being found by prospectors using drywashers in Trinity Canyon which is about 10 miles north of Lovelock. There is no recorded production for the area.

Unionville District

The placers here were some of the earliest known in the state. The first to work them were the Spanish long before the rush. Some of the gravels were very rich, and some of the early miners recovered as much as two ounces a day using hand methods. One report states that two men using a rocker got 15 ounces in a day and a half in the 1870's. In 1875 there were over a hundred and fifty men working in Indian Canyon alone. Since so much of the gold was mined by the Chinese there are no accurate records available as to the amount of production from this district. This district is also known as the Buena Vista district. Mark Twain was one of the first to join the rush to this area and may have acquired some of his inspiration for his book on the miner's life, "Roughing It," from his time spent here. The most productive areas have been in Buena Vista Canyon, Congress Canyon, and Indian Canyon. Unionville was Humboldt County seat from 1863 to 1873 and has the oldest schoolhouse in the state, which is now a museum. To reach Unionville go west on Highway 80 about 1 mile from Mill City. Take the side road going south about 16 miles to a side road on your right. Unionville is about 3.5 miles up this road.

Willow Creek

Placers have been worked in recent times in Willow Creek. Estimated production is set at about 5,000 ounces. Willow Creek is located about 12 miles southeast of Mill City and flows into Dun Glen Creek.

Agate-Chalcedony-Opal

To reach this collecting area go west on Highway 48 about 14 miles to the Eagle Pitcher Mine Road. Stay on the road to the mine for about 6 miles. Search on both sides of the road for specimens. The collecting area is about a mile in diameter.

Agate Geodes

Take Highway 50 south from Highway 80 near Mill City for about 10 miles to the area around Star Peak. Search along the eastern slope of Star Peak.

Fossils

Take Highway 80 north from Lovelock for about 7 miles to a paved road going east. Follow this road for about 14 miles to the Dago Pass turnoff; this is a good dirt road. Stay on this road for about 6.5 miles to the east side of a low summit in the Humboldt Mountains. Search on both sides of the road up to the base of the hills and in the flat.

Agate-Jasper-Opal

Go west on Highway 48 from Lovelock about 2 miles to a dirt road heading southwest. Take this road about 13 miles to a branch road leading into the Trinity Mountains. Follow this road for about 3.5 miles to the collecting area.

Calcite-Garnet-Scheelite

Take Highway 80 south from Lovelock about 14 miles to the Nightingale exit and go west about 10 miles to a fork. Take the fork to the left, going southwest about 18 miles to another fork. Take the right hand fork going northwest. It is about 9 miles down this road to an old mining camp; search in the old mine dumps.

Apache Tears-Sunstone

The area south of the old mining camp of Placeritas has produced some good specimens of Apache tears and sunstone. Follow directions to the Placeritas mining district.

Scossa Ghost Town

Rich jewelry rock was found here in 1906 by the Scossa brothers, and by 1907 a small town had grown up to serve the miners working in the nearby hills. Only a few buildings remain to mark the site of this once-active community. To reach Scossa go west from Jungo on Highway 49 about 2 miles to a side road going south. Follow this road about 17 miles to the site of Scossa.

Treasure Tales

This story begins when the Humboldt Sink still contained rich grassland. The first wagon trains were able to graze their stock there to get them ready for the rugged desert crossing that lay ahead. While their stock was fattening up on the grass the pioneers would search the mountains for game. During one of these hunting trips a man picked up what he thought was lead ore and brought it back to camp. He threw the ore in a bucket in his wagon and forgot about it. Years later when he was living in San Francisco a friend of his who knew a little about mining saw one of the pieces of ore being used for a door stop. He asked the man if he could have it assayed. The pioneer, now an old man, agreed. The assay turned out rich in silver and lead. The old man was not able to travel and a few years later he passed away. His grandsons discovered documents the old man had made up showing where he had found the ore. In 1909 the grandsons stepped off the stage in Humboldt City. They headed almost at once in search of the lode. The young men spent two weeks looking for the location, then returned to Humboldt City. The men did find a piece of rich ore that matched their grandfather's ore. They said they had found it in an abandoned miners camp site but didn't know where it had come from. According to the story, the grandfather had gone in an easterly direction and was gone for only a few hours from his camp site in the sink. The young men had gone in a westerly direction when they first left Humboldt City.

91

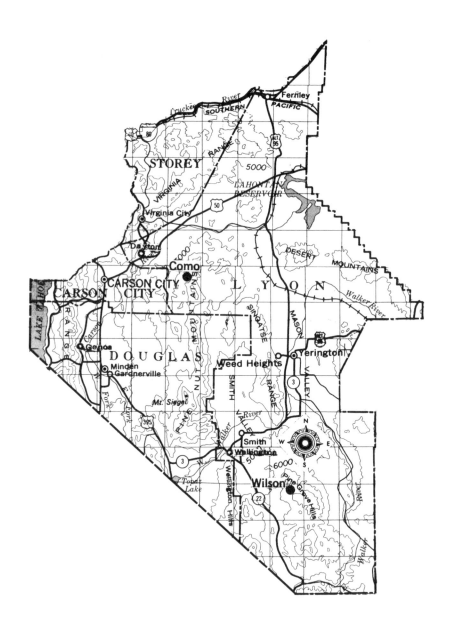

Storey County

The history of gold in Storey County is the Comstock Lode. It was the discoveries here that brought in the prospectors and miners and led to the search for minerals throughout the state. From 1859 to 1939 Storey County led the state in gross production of minerals. It was not surpassed until 1940, when White Pine County took the lead away. Some of the early records were destroyed in the great fire in Virginia City in 1875. This is a small county but it contains some good gem collecting areas as well as a treasure tale or two and, of course, the town of Virginia City.

Comstock District

This district is located in the area around Virginia City and includes Gold Hill. The ores occur in a mineralized shear zone several hundred feet wide and 13,000 feet long. The country rock is make up of Triassic sedimentary and volcanic rocks, Jurassic quartz monzonite, and Tertiary volcanic rocks. This is primarily a silver mining district and has produced more than 9 million ounces of that metal over the years. Silver made them rich, but it was placer gold that first brought them to the area. The placer gold had originated from the ores that would become known as the Comstock Lode, and it was not until later that the miners would learn that they were throwing more treasure away than they were keeping in their pans and rockers. There are no records available on the amount of placer gold mined, but the lode production is set at 8,500,000 ounces. Take Highway 79 east from Virginia City and try panning along the Carson River. You can still get some color here. Try your drywasher in the canyons.

Cinnabar-Opal

Go north on Highway 17 at Virginia City and take Nevada 45 northeast at the grade. Follow Nevada 45 across Five Mile Flat and search along canyon walls and floor for collectables.

Quartz Crystals-Silver Specimens

Try searching in the mine dumps in the Comstock mining district for quartz crystals, calcite, gypsum, and silver minerals.

Virginia City Ghost Town

To call Virginia City a ghost town is not really being accurate. It is now a tourist attraction and the historic buildings now house souvenir shops and the like. It can be a disappointment if you are a serious ghost town searcher. However, some of the old homes that have been restored are worth seeing. Take Highway 50 northeast from Carson City to the Highway 17 turnoff and follow it to Virginia City; it is about 16 miles.

Treasure Tales

Virginia City is named for prospector James Fennimore, who was known as "Old Virginny." Fennimore is known to have tipped a few in his day and was a very poor businessman. There is a story told about how he made a discovery of a nice gold-bearing vein on what was then known as Sun Mountain and is now named Mount Davidson. He mined the ore himself for a spell, just taking out enough to fortify himself each night at one of the local saloons. Some mining speculators felt that his claim was worthwhile and offered to buy him out. He sold the claim for a bag of gold coins worth $7,500 and a bottle of whiskey. Fennimore sat down and downed the quart of whiskey on the spot. He figured he'd hide the bag of coins and go to the saloon and celebrate. He must have really hung one on because when he sobered up a few days later he couldn't remember where he had hidden the gold. He knew it was somewhere near his mine and that he had buried it under a big rock. He never did find the gold and it could still be there today.

Washoe County

This is a fairly large county that forms the northwestern boundary with California. Washoe County has not been a large producer of gold over the years, but there are several spots that have been productive. The only area with any recorded production has been the Olinghouse district. There are some good collecting areas for the rockhound here. Also, this is a good location for the treasure hunter as there are several lost treasure stories told here.

Galena Creek

Some prospectors report that you can get a little color from Galena Creek. To reach this area go south on Highway 395 from Reno about 10 miles to the junction with Highway 27. Go west on Highway 27 about 6 or 7 miles to where the road crosses the creek. Steamboat Creek south of here also has a little gold.

Jumbo District

This district is also known as the West Comstock district and is located just across the county line from Storey County, west of Gold Hill. Some small scale lode and placer mining has been done here, but there is no recorded production for the district.

Little Valley

Little Valley occupies a trough that runs north and south for a distance of about 4 miles. It is about 5 miles southwest of the old camp of Franktown. Franktown can be reached by going north on Highway 395 about 8.5 miles from Carson City to the Franktown Road; it is just a little over a mile to the site. Although there are no recorded production figures this is known to have been a profitable placer mining area. It is located at about 6,500 feet in the Sierras. There are tailing piles all over that attest to the amount of work done here. This is one of the areas where the gold is found in a Tertiary river channel. This ancient channel consists of well-rounded pebbles and boulders of quartzite, schist, granodiorite, and other rocks, and in places is overlaid by flows of rhyolite and andesite. There have been some reports of rich pockets found here in the past. One pocket is said to have contained $60,000 in gold at the old price of around $20 an ounce. Water is available year round here from Franktown Creek.

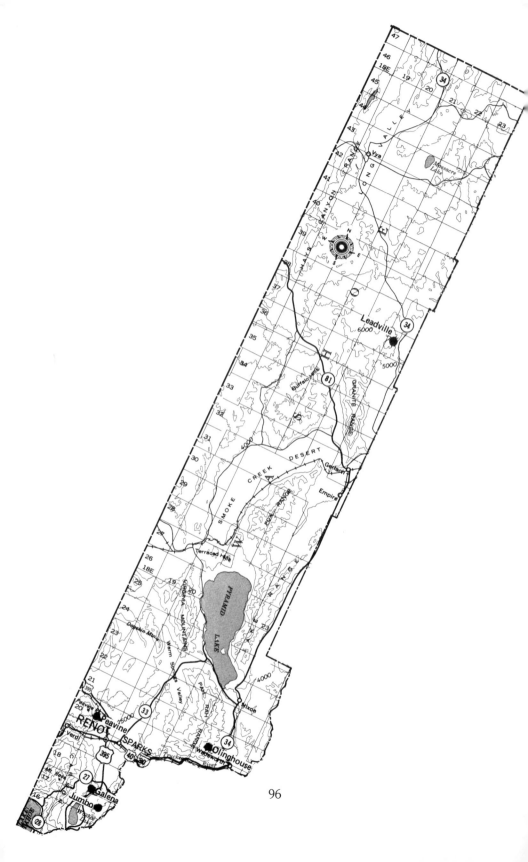

Olinghouse District

This is the only district in the county for which any production figures are available. The recorded production for the district is set at 36,000 ounces, all from lode. The ore occurs in zones of altered Tertiary andesite adjacent to intrusive rhyolite. There has also been a fair amount of placer gold found here. To reach the area go east on Highway 80 from Reno to Wadsworth. Take Highway 34 north on this road a little over 2 miles to a side road going west. It is 6 miles down this road to the site of Olinghouse. This is also known as the White Horse district and is on the east slope of the Wilcox Mountains. The placers were worked as early as 1860 and were quite productive before 1900. The deposits are in the ravines running into Olinghouse Canyon. The gravels are cemented in some parts and moist a few inches down and can be hard to drywash.

Peavine District

This is also known as the Reno or Crystal Peak district. The district is located on the northern slope of Peavine Mountain about 6 miles northwest of the city of Reno. The lode deposits were discovered in 1863, and the first rail shipment over the Sierras contained ore from this district shipped to Sacramento in 1866. The main placering area is in the ravine and the most productive years were from 1876 to 1881.

Obsidian-Opalized Wood

Go north from Gerlach on Highway 81 a little over 30 miles to a dirt side road on your right. Take this road about 8 to 9 miles to a side road going northeast. Take this road to the collecting area; it covers an area of about 2 miles.

Agate-Apache Tears-Jasper-Opalized Wood

Go north on Highway 34 from Gerlach about 45 miles to Petrified Stump State Park. About 8 miles past the park search both sides of the road for about 3 miles.

Agate-Chalcedony-Jasper-Opal-Wood-Rhyolite

Just south of Vya take Nevada 8A off Highway 34. Follow it 22 miles to a side road on your right. About 3 miles down this road search both sides for rhyolite specimens. Another mile or so down the road you will come to a side road going north to an opal and chalcedony collecting area. Staying on the road for a total of 7 miles will bring you to Wall Canyon. Prospect in the canyon for wood, opal, and jasper.

Cinnabar in Opal

Go south on Highway 395 from Reno. Just south of the junction with Highway 27 you will come to a side road going west into the geyser areas. Search around the geysers for cinnabar in opal.

Leadville Ghost Town

There are still a few buildings standing here as well as the ruins of the old mill. The boom here lasted from 1910 until the early 1920's. This could be a good metal detecting site. Some of the mines are still under claim and some work still goes on, so check before you hunt.

Treasure Tales

In 1870, a Central Pacific train was held up by bandits near the town of Verdi, west of Reno. The robbers had been tipped off that the train would be carrying a large number of gold coins. The bandits got nearly $60,000 in loot from the robbery. They buried their partner's share near an old mine near the railroad tracks and took off. The bandits were captured and their share of the loot recovered. They were sent to prison, but they never revealed where they had buried the other gold coins.

Two chests of gold nuggets and coins are said to be buried near the cliff looking over the southern part of Pyramid Lake. The gold is said to have belonged to some Chinese miners who were ambushed by Indians. The Indians had no use for the gold, so they buried it after the massacre. There have been several searches for the gold but it never has been found.

White Pine County

Up until 1869 what is now Lander County was a part of White Pine County. The early recorded figures only go back to 1865, and the production prior to that is not known. That there was quite a bit of placer gold taken out in the 1850's and early 1860's by prospectors we can be sure. The largest mining operation in the state for many years was carried on by the Nevada Consolidated Copper Company in the Ely district. By the early 1960's the Ely district had produced $900,000,000 worth of copper and $80,000,000 in gold and silver and other minerals. One of the most popular spots for modern prospectors is located here, as well as some good sites for the treasure hunter to use his metal detector, and a couple of gem sites for the rockhound.

Bald Mountain District

This district is also known as the Joy district and is located on the western slope of the Ruby Range of mountains in the northwestern part of the county. This is another district that was worked by the Chinese miners early in its history. The placer deposits are in Water Canyon and the ravines leading into it. The most productive area has been in the lower end of Water Canyon. The richest gravels are found on or near bedrock, but the overburden also carried values. Some fair-sized nuggets have been found here. The mining season is short here as the elevation is 7,400 feet and it gets cold. The scarcity of water has kept any large scale operations from taking place. To reach this area go east on Highway 50 from Eureka about 15 miles to a paved road going north. Follow this road about 38 miles to a side road heading east. Take this road 5 miles to a side road going north. It is about 8 miles up this road to the site of Joy.

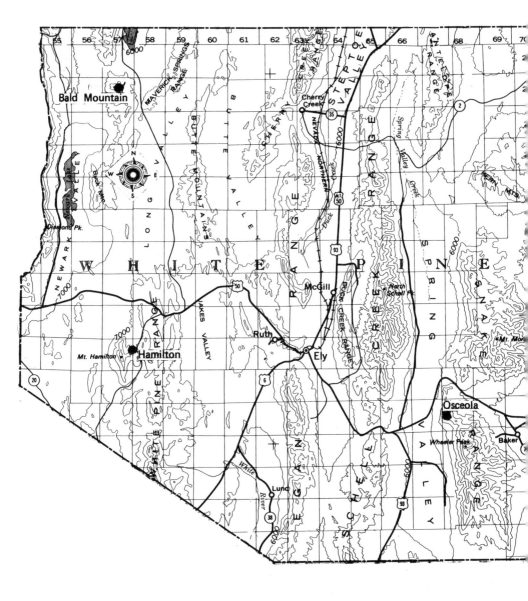

Cherry Creek District

Most of the gold mined here has been as a by-product of silver-lead ores. The veins occur in Cambrian quartzite near small quartz monzonite and diabase intrusives. Production is set at 36,200 ounces, all from lode. There are reports of small amounts of placer gold being found in Bagan Canyon. To reach this district go north on Highway 93 from Ely for 45 miles to Nevada 35. It is about 9 miles down this road to Cherry Creek.

Osceola District

This district is located on the western flank of the Snake Range of mountains in southern White Pine County. It is the best placering area in the county and some of the largest nuggets found in the state come from this district. The lode deposits were discovered in 1872 and the ore is gold-bearing quartz in mineralized sheeted zones and scattered zones in Cambrian quartzite. The placer deposits are spread over a wide area. The most productive placers are in Dry Gulch and Grup Gulch near the town of Osceola, and in Weaver Creek, Mary Ann Canyon, and the Summit placer diggings. Production for the district is set at 91,600 ounces placer and 40,100 ounces lode. What is probably the largest nugget ever found in the state was discovered in 1878. It weighed 24 pounds. Various methods have been used over the years to work the deposits here, including hydraulics in Dry Gulch. To reach this area take Highway 50-93 south from Ely. Continue on Highway 50 past the junction of Highway 93 for 8 miles to a side road on your right. It is about 4.5 miles down this dirt road to the site of Osceola.

Wonderstone

From Ely go west on Highway 50 for about 40 miles to Little Antelope Summit. Search on the north side of the highway in this area for rhyolite wonderstone.

Garnet in Rhyolite

Go west on Highway 50 from Ely for about 5 miles to a side road going north; this is a good gravel road. Follow this road for less than a mile into a canyon. Prospect in rhyolite for garnets.

Fluorescent Calcite

Take Highway 50 southeast for about 28 miles to Conners Pass. Look in roadcut next to highway for fluorescent calcite.

Hamilton Ghost Town

In 1869 Hamilton was a booming city of 15,000, and is completely deserted today. There are still buildings standing as a reminder of what can happen in the boom and bust of mining. It is well worth seeing and a good treasure hunter might make some good finds with his metal detector here. To reach the site of Hamilton go west on Highway 50 for about 37 miles from Ely to a side road going south. Hamilton is about 10 miles down this road.

Treasure Tale

There is a lost mine story that is set in the Ruby Mountains. A pioneer family who had a cabin on the southern tip of Ruby Lake was murdered by a band of renegade Indians around 1878. It was almost 30 years later before their remains were found. A wagon holding almost two tons of rich gold ore was found with the remains of the family, but no clue to its source was found. People have searched for years in the region for the Lost Pioneer Mine but it has never been found.

GEOLOGY OF PLACER DEPOSITS

Geologic evidence shows that there was an abundance of water in Nevada at one time. This is proven by the presence of fossil shells in isolated patches throughout a large portion of the state. The evidence of water action in the formation of gold placers in the state is shown in some of the well-rounded pebbles and gold particles. The placer material was transported by gravity, rains, and melting snows into the beds of ravines or streams that are now dry. In the desert areas of Nevada, where disintegration of the rocks produces masses of loose material, mostly angular, which is washed to lower land by the infrequent but heavy rains that occur, the gold is found in loose alluvium or wash where there are no rounded pebbles or rocks. The placer gold concentrations in these more arid regions may also be erratic, as the gold was deposited largely by intermittent rather than by constant stream action. When cloudbursts occur in the arid regions, the flow of water lasts for only a few hours. There is no time to soften the underlying earth except for a few inches and, consequently, the gold is found in narrow streaks of limited length and depth. The pay streaks are irregular in occurrence and have to be located by hit-and-miss prospecting. The gravels in the dry placers are usually a mixture of fine and coarse material that shows little evidence of stratification.

In several locations in Nevada, placers occur that were formed by ancient Tertiary streams. No doubt some of these ancient channels contain gold and are of economic importance. A careful geologic study may lead to the discovery of ancient placers in the state that have escaped observation. The Tertiary streams of the Mother Lode region on the western slope of the Sierra Nevadas in California were very rich in places. A particularly large and well-defined ancient Tertiary stream that had an east-west course has been traced from the northeast corner of Lake Tahoe to a point about 4 miles north of Carson City. This river is believed to have flowed eastward, but complete evidence is missing. There are several places in the state where Tertiary placer gravels have been worked, some of the more productive being at Genoa, 15 miles south of Little Valley, east of Genoa in the Mount Siegal district and, a little farther east, Smith Valley and Yerington placers.

In Nevada, with the exception of the Tertiary river channels, most of the placer gold came from nearby lode deposits. Most of the gold you will find here has no doubt traveled less than 10 miles. The sediments that fill the valleys in the gold districts also contain large amounts of gold, but the gold is too scattered to be mined economically.

In order to have the best chance at finding that golden dream you are seeking, you need to have some knowledge of placer deposits. A lot of our information comes from the early miners and prospectors who climbed, dug into, and checked out most every mountain, canyon, stream, river, and creek in our great state. This is still the best method, as geologists admit that even today they don't know everything there is to know about the remaining rich gravel deposits. In a Mineral Information Service Bulletin put out by the State of California Division of Mines and Geology they stated, "The geologic history and structure of the buried channels are so complex that the best of engineers have been baffled by them. Fragmentary benches and segments of rich gravel deposits which still rest in positions completely hidden from the surface, or even from the underground passages which enter into the lower main channels, afford alluring possibilities to the geologist as well as the prospector." So they are telling us that there is still a lot of gold out there, and you've as good a chance at finding it as any geologist.

The number one thing to keep in mind is that most all areas have been prospected at one time or the other. Don't waste a lot of time in areas that have not proven to be productive in the past. Search the areas that are known to be gold bearing, and take advantage of the knowledge gained by those who went before you.

There are several types of placer deposits which are classified here to indicate how they were first formed. The basic placers are:

1. Residual placers or "Seam Diggens."
2. Eluvial or hillside placers, representing transitional creep from residual deposits to stream gravels.
3. Bajada placers, a name given to a peculiar type of desert or dry placers.
4. Stream placers which have been sorted and resorted and are simple and well merged.
5. Glacier-stream placers which are, for the most part, profitless.
6. Eolian placers, or local concentrations caused by the removal of lighter materials by the wind.
7. Marine or beach placers.

Of the seven types, the stream placers are the most important. They have been the source of most of the placer gold mined. Stream placers consist of sands and gravels sorted by the action of running water. If they have undergone several periods of erosion and have been resorted, the greater the concentration of the heavier minerals. Deposits by streams include those of both present and ancient times, whether they form well-defined channels or are left merely as benches. All bench placers, when first laid down, were stream placers similar to those of the present stream deposits. If not reworked by later erosion, they are left as terraces or benches on the sides of the valley cut by the present stream. These deposits are called bench gravels. In order to understand stream placers, streams themselves must be studied in regard to their habit, history and character.

Residual placers are formed when the gold is released from its source and the encasing material broken down. This is most effectively done by long continued surface weathering. Disintegration is accomplished by persistent and powerful geologic conditions which affect the mechanical breaking down of the rock and the chemical decay of the minerals. The surface of a gold-bearing ore body is enriched during this process of rock disintegration because some of the softer and more soluble parts of the rock are carried away by erosion.

After gold is released from its bedrock encasement by rock decay and weathering it begins to creep down the hillside and may be washed down rivulets and gullies and into the stream beds. On its way down the hillside the gold is sometimes concentrated in sufficient value to warrant mining. These deposits are classified as eluvial deposits.

It is a common fallacy of some prospectors to attribute the forming of some placer deposits to the action of glaciers. Since it is the habit of glaciers to scrape off loose soil and debris but not to sort it, ice is ineffective in the concentration of metals. The streams issuing from the melting ice may sometimes be effective enough in sorting to create a deposit.

"Bajada" is a Spanish word for slope and is used to identify a confluent alluvial fan along the base of a mountain range. The total production of gold from bajada placers is small compared to other placer workings due to the less efficient dry washing methods used in the past. The forming of a bajada placer is basically similar to a stream placer except as its condition by the climate and topography of the arid region in which it occurs. The bulk of the gold that has been released from its matrix as it travels from the lode outcrop to the bajada slope is deposited on the slope close to the mountain range. The gold is dropped along the lag line, which is the contact of the basin fill with the bedrock. Although the heaviest concentration of gold will be on bedrock, bulk concentration does not occur as in a stream deposit. Since a certain percentage of gold is still locked in its matrix, there is a strong tendency for less gold to reach bedrock and for more to remain distributed throughout the deposit than in the case of stream gravels.

Most of the Eolian placers of the desert are as a result of the bajada being enriched on the surface by wind action on the lighter materials.

There are several things that occur to preserve a placer deposit. Since streams are constantly changing their position, fragments of their deposits are left isolated. For example, the benches and terraces that are left at different intervals when a stream is cutting a deeper channel. These deposits that are left will eventually be eroded away unless something protects them. Burial is the most effective way a placer may be preserved. Mostly, when the name "buried channel" is given to a placer it is one where a stream has been covered by lavas, mud-flows, ash falls, etc. There are other ways by which they may be buried, such as:

1. By covering with landslide material.
2. By covering with gravel.
3. By covering with lake deposits.
4. By covering with gravel when the stream is choked.
5. By covering with gravel when the stream course is lowered below the general base-level of erosion.
6. By the covering of older stream courses with alluvial fan material, as conditions favorable to stream existence fail.
7. By covering with glacial till.
8. By covering with beach placers with marine sediments as the coast is submerged and elevated.

The gravel content of a placer may become firmly cemented when it is infiltrated by mineral matter such as lime and iron carbonate or silica. The older the placer the more likely this is to occur. These cemented gravels sometimes are very hard to break down. This is why some of the tailings in the old mines are profitable to work. The cemented gravels sometimes were never completely broken down as they traveled through the sluice boxes, and the gold was redeposited in the present stream bed.

The gold found in placers originally came from veins and other mineralized zones in bedrock when the gold was released from its rock matrix by weathering and disintegration. Many times the sources of the gold in a placer would not be a deposit that could be mined at a profit, but the richer deposits usually indicate a comparatively rich source. Sometimes a rich placer will be developed when several low grade veins feed it over a very long time. The richest placers are created when there is reconcentration from older gold-bearing gravels. For the most part, the original source of the gold is not far from the place where it was first deposited after being carried by running water.

The mineral grains that are very heavy and resistant to mechanical and chemical destruction will be found with the gold in placer deposits. These are what prospectors and miners call black sand. The black sand is generally principally magnetite, but some of the other minerals you will find in your sluice box are: garnet, zircon, hemitite pyrite ("Fool's Gold"), chromite, platinum, cinnabar, tungsten minerals, titanium minerals, as well as diamonds. You'll find a lot of other things such as quicksilver, metallic copper, amalgam, nails, buck-shot, B.B.s, bullets, and what have you.

The very high specific gravity of gold (six or seven times that of quartz, and increasing to nine times under water) is what causes the gold to work its way to the bedrock or false bedrock, or any point where it can go no further. Once it is trapped on bedrock, the stream has great difficulty picking it up again. The specific gravity of gold is 19; that is, it weighs 19 times as much as an equal amount of water to its mass.

AVERAGE SPECIFIC GRAVITY OF SOME MINERALS

MICA	2.3
FELDSPAR	2.5
QUARTZ	2.7
HORNBLENDE	3.2
GARNET	3.5
CORUNDUM	4.0
MAGNETITE	5.2
SILVER	7.5
GOLD	19.2

Due to its insolubility, the finest particles of gold are preserved. A piece of gold worth less than a penny can easily be recognized in a pan. Since gold is so malleable it will be hammered into different shapes by stones as they tumble along in the stream. It will not be welded together to form larger nuggets as some people believe. Particles of gold may be broken down, however, from another piece. Geologists have shown that the largest masses of gold come from lodes and not placers. The more rounded and flattened nuggets that you find have probably been in the stream for a longer amount of time and have taken more knocking around than the ones that show the original crystalline form. The crystalline nuggets are known as coarse gold and probably have not traveled far from the source in the free state.

The gold found in the more ancient placers has a higher degree of fineness than that from a nearby source. This may be due to the removal of alloyed silver by the dissolving action of the water.

The accumulation of gold in an important placer deposit is not just pure coincidence but is the result of some fortunate circumstances. In areas where nature has provided extensive mineralization, rapid rock decay, and well-developed stream patterns, there is the opportunity for large amounts of gold placer to be formed. Basically what happens is fairly simple: in areas where the gold has been deposited, the power of the stream has become insufficient to carry off the particles of gold that have settled. How rich the deposit is will depend on how complete the loss of transporting power is, as well as the ability of the bedrock to hold the deposited gold, plus the general relationship of the gold sources to the stream.

When a stream is eroding, the materials in reach of its activity are constantly moved downstream. During this movement, a constant sorting is taking place, which causes a concentration of the heavier particles. Dispositon then takes place in the stream when the velocity is decreased, either by changes in volume or grade. When this happens, the gold is laid down with the other sediments. Sometimes the placer gold may be trapped in irregularities in the bedrock without considerable detrital material being trapped with it, but extensive placers—as a rule—are not formed by irregularities in the bedrock alone. When the bed of the stream is the actual rock floor of the valley, this is true bedrock. When the gravels become covered with volcanic or other materials, the stream will flow over this new floor, making deposits on what is known as false bedrock. So you can see that an area may contain two or more layers of gold-bearing gravels. An easy way to see how a stream lays down these various layers is to study areas where roadcuts have exposed ancient stream deposits and also in the canyons where the benches can be seen.

A smooth, hard bedrock is a very poor one to develop a good placer deposit. The bedrock formations that are highly decomposed and possess cracks and crevices are good, and those of a clayey shistose nature are rated excellent in their ability to trap the gold.

To give you an idea of the carrying power of a stream, here are some figures on the size of material carried by a stream flowing at different velocities:

3 in. per second—0.170 mph, will just begin to work on fine clay.

6 in. per second—0.340 mph, will lift fine sand.

8 in. per second—0.4545 mph, will lift sand as coarse as linseed.

10 in. per second—0.5 mph, will lift gravel the size of peas.

12 in. per second—0.6819 mph, will roll along gravel the size of beans.

24 in. per second—1.3638 mph, will roll along rounded pebbles 1 inch in diameter.

3 ft. per second—2.045 mph, will sweep along slippery, angular stones the size of hens' eggs.

As far as grade is concerned, a grade ranging from 30 to approximately 100 feet per mile will favor the disposition of gold. Anywhere that the grade is greater than that, such as in mountain streams in narrow canyons, will not be a good source of placer deposits. When a stream leaves its mountain canyons and enters a more level country or a still body of water, the material carried by the stream is deposited in the form of a fan or a delta. At the apex of a fan or delta, the fine gold will be deposited and may never reach bedrock. The steering action that occurs in the rugged mountains during times of floods, which permits gold to reach bedrock, does not take place in the delta.

To sum things up, remember that the gold is heavy, heavier than most of the other material in the stream bed, and it will drop anywhere the flow or grade changes, causing the stream to slow down or lose its carrying power. These are the places you want to search. Each rainy season will bring new gold down from the hillsides into the stream bed. Another thing to keep in mind is the fact that the early miners were working deposits that had had thousands of years to develop. Try to find material that has not been worked before, or try to reach the hard-to-get-to areas where the chances are that the gravels have not been worked as much. The most important thing to keep in mind is that "GOLD IS WHERE YOU FIND IT."

GLOSSARY

ALLOY. A solid solution of two or more minerals.

ALLUVIAL. Loose gravel, soil, or mud, deposited by water.

AMALGAM. Normally a physical alloy of mercury with gold or silver.

ARRASTRE. A circle of stones where ore was crushed during the early days of gold mining; a primitive but effective method of separating gold from quartz.

ASSAY. To evaluate the quantity and quality of the minerals in an ore.

BAR. A name given to the sandbars and rock and gravel bars found in rivers, primarily when they are gold-bearing.

BENCHES. All kinds of rock or gravels shaped like terraces or steps. Bench placers are found on the canyon walls above the present stream beds.

BLACK SAND. This is usually made up of magnetite, tourmaline, ilmenite, chromite, and cassiterite and is found in rivers, beaches, sluice boxes, and pans. Black sand will usually be found with gold, but gold is not always found with black sand.

COLOR. Any amount of gold found in a prospector's pan after a sample of dirt has been panned.

DIGGINGS. A claim or place being worked.

DIORITE. A granular, crystalline, igneous rock in which gold sometimes occurs.

DREDGING. A method of vacuuming gold-bearing gravels from river or stream bottoms.

DRIFT. A horizontal tunnel following a vein or gold-bearing gravels.

DRY WASHER. A machine which separates gold from gravels by the flow of forced air.

FLOAT. Loose pieces of ore broken off a vein outcropping. Prospectors will follow the float to its source to locate a lode.

GLORY HOLE. A small but very rich deposit of gold ore.

GRAVEL BENCHES. Gravel deposits left on canyon walls through stream erosion.

GULCH. A Small canyon or ravine.

HARDROCK MINING. Another term for lode mining.

HEADFRAME. The support structure located at the entrance of a mine over a shaft. Used for hoisting.

109

HYDRAULIC MINING. A very destructive and now outlawed form of gold mining used in the Gold Rush. Giant hoses were used to force great streams of water onto canyon walls containing gold-bearing gravels. The walls were washed away into sluice boxes where the gold was then picked out.

IRON PYRITE. A common mineral consisting of iron disulfide which has a pale, brass-yellow color and a brilliant metallic luster. (Also called ''Fool's Gold.'')

LODE. A vein of gold mined either through a tunnel or a shaft.

MATRIX. The material in which the gold is found.

PLACER. Free-occurring gold which is usually found in stream and river gravels.

POCKET. A rich deposit of gold occurring in a vein or in gravels.

POKE. A leather pouch used by old-time miners to hold their gold.

QUARTZ. A common mineral, consisting of silicon dioxide, that often contains gold or silver.

RETORT. A device used to separate gold from mercury.

RICH FLOAT. Gold-bearing rocks worked loose from a lode.

ROCKER. A device used by the early miners during the Gold Rush. This was a sluice box mounted on rockers with a hopper on the top to classify the material. Gravels were shoveled into the hopper; then water was poured on top, washing the gold-bearing material down over the riffles while the hopper was rocked. The rocking helped the gold to settle.

SCHIST. A crystalline rock which is easily split apart.

SLUICE BOX. A trough with obstructions to trap gold used in continuously moving water.

STAMP MILL. A machine used to crush ore.

SULFIDE. A compound of sulfur and any other metal.

TAILINGS. The material thrown out when ore is processed. The tailings from the early mines, where the miners were sometimes very careless, have produced significant amounts of gold and other valuable minerals.

TERRACE DEPOSITS. Gravel benches high on canyon walls.

WALL. The rock on either side of a vein.

WIRE GOLD. Gold thinly laced through rock.